T0275940

Data-Centric Systems and Applications

For further volumes:
www.springer.com/series/5258

Stefano Ceri · Alessandro Bozzon ·
Marco Brambilla · Emanuele Della Valle ·
Piero Fraternali · Silvia Quarteroni

Web
Information
Retrieval

 Springer

Stefano Ceri
Dipartimento di Elettronica e Informazione
Politecnico di Milano
Milan, Italy

Emanuele Della Valle
Dipartimento di Elettronica e Informazione
Politecnico di Milano
Milan, Italy

Alessandro Bozzon
Dipartimento di Elettronica e Informazione
Politecnico di Milano
Milan, Italy

Piero Fraternali
Dipartimento di Elettronica e Informazione
Politecnico di Milano
Milan, Italy

Marco Brambilla
Dipartimento di Elettronica e Informazione
Politecnico di Milano
Milan, Italy

Silvia Quarteroni
Dipartimento di Elettronica e Informazione
Politecnico di Milano
Milan, Italy

ISBN 978-3-642-43118-0   ISBN 978-3-642-39314-3 (eBook)
DOI 10.1007/978-3-642-39314-3
Springer Heidelberg New York Dordrecht London

ACM Computing Classification (1998): H.3, I.2, G.3

Printed on acid-free paper

Springer is part of Springer Science+Business Media (www.springer.com)

Preface

While information retrieval was developed within the librarians' community well before the use of computers, its importance boosted at the turn of the century, with the diffusion of the World Wide Web. Big players in the computer industry, such as Google and Yahoo!, were the primary contributors of a technology for fast access to Web information. Searching capabilities are now integrated in most information systems, ranging from business management software and customer relationship systems to social networks and mobile phone applications. The technology for searching the Web is thus an important ingredient of computer science education that should be offered at both the bachelor and master levels, and is a topic of great interest for the wide community of computer science researchers and practitioners who wish to continuously educate themselves.

Contents

This book consists of three parts.

- The first part addresses the *principles of information retrieval*. It describes the classic metrics of information retrieval (such as precision and relevance), and then the methods for processing and indexing textual information, the models for answering queries (such as the binary, vector space, and probabilistic models), the classification and clustering of documents, and finally the processing of natural language for search. The purpose of Part I is to provide a systematic and condensed description of information retrieval before focusing on its application to the Web.
- The second part addresses the *foundational aspects of Web information retrieval*. It discusses the general architecture of search engines, focusing on the crawling and indexing processes, and then describes link analysis methods (and specifically PageRank and HITS). It then addresses recommendation and diversification as two important aspects of search results presentation and finally discusses advertising in search, the main fuel of search industry, as it contributes to most of the revenues of search engine companies.

- The third part of the book describes *advanced aspects of Web search*. Each chapter provides an up-to-date survey on current Web research directions, can be read autonomously, and reflects research activities performed by some of the authors in the last five years. We describe how data is published on the Web in a way to provide usable information for search engines. We then address meta-search and multi-domain search, two approaches for search engine integration; semantic search, an important direction for improved query understanding and result presentation which is becoming very popular; and search in the context of multimedia data, including audio and video files. We then illustrate the various ways for building expressive search interfaces, and finally we address human computation and crowdsearching, which consist of complementing search results with human interactions, as an important direction of development.

Educational Use

This book covers the needs of a short (3–5 credit) course on information retrieval. It is focused on the Web, but it starts with Web-independent foundational aspects that should be known as required background; therefore, the book is self-contained and does not require the student to have prior background. It can also be used in the context of classic (5–10 credit) courses on database management, thus allowing the instructor to cover not only structured data, but also unstructured data, whose importance is growing. This trend should be reflected in computer science education and curricula.

When we first offered a class on Web information retrieval five years ago, we could not find a textbook to match our needs. Many textbooks address information retrieval in the pre-Web era, so they are focused on general information retrieval methods rather than Web-specific aspects. Other books include some of the content that we focus on, however dispersed in a much broader text and as such difficult to use in the context of a short course. Thus, we believe that this book will satisfy the requirements of many of our colleagues.

The book is complemented by a set of author slides that instructors will be able to download from the Search Computing website, www.search-computing.org.

Milan, Italy   Stefano Ceri
                                                                                Alessandro Bozzon
                                                                                  Marco Brambilla
                                                                            Emanuele Della Valle
                                                                                   Piero Fraternali
                                                                                Silvia Quarteroni

Acknowledgements

The authors' interest in Web information retrieval as a research group was mainly motivated by the Search Computing (SeCo) project, funded by the European Research Council as an Advanced Grant (Nov. 2008–Oct. 2013). The aim of the project is to build concepts, algorithms, tools, and technologies to support complex Web queries whose answers cannot be gathered through a conventional "page-based" search. Some of the research topics discussed in Part III of this book were inspired by our research in the SeCo project.

Three books published by Springer-Verlag (*Search Computing: Challenges and Directions*, LNCS 5950, 2010; *Search Computing: Trends and Developments*, LNCS 6585, 2011; and *Search Computing: Broadening Web Search*, LNCS 7358, 2013) provide deep insight into the SeCo project's results; we recommend these books to the interested reader. Many other project outcomes are available at the website www.search-computing.org. This book, which will be in print in the Fall of 2013, can be considered as the SeCo project's final result.

In 2008, with the start of the SeCo project, we also began to deliver courses on Web information retrieval at Politecnico di Milano, dedicated to master and Ph.D. students (initially entitled *Advanced Topics in Information Management* and then *Search Computing*). We would like to acknowledge the contributions of the many students and colleagues who actively participated in the various course editions and in the SeCo project.

Contents

Part I
Principles of Information Retrieval

Chapter 1
An Introduction to Information Retrieval

Abstract Information retrieval is a discipline that deals with the representation, storage, organization, and access to information items. The goal of information retrieval is to obtain information that might be useful or relevant to the user: library card cabinets are a "traditional" information retrieval system, and, in some sense, even searching for a visiting card in your pocket to find out a colleague's contact details might be considered as an information retrieval task. In this chapter we introduce information retrieval as a scientific discipline, providing a formal characterization centered on the notion of relevance. We touch on some of its challenges and classic applications and then dedicate a section to its main evaluation criteria: precision and recall.

1.1 What Is Information Retrieval?

Information retrieval (often abbreviated as IR) is an ancient discipline. For approximately 4,000 years, mankind has organized information for later retrieval and usage: ancient Romans and Greeks recorded information on papyrus scrolls, some of which had tags attached containing a short summary in order to save time when searching for them. Tables of contents first appeared in Greek scrolls during the second century B.C.

The earliest representative of computerized document repositories for search was the Cornell SMART System, developed in the 1960s (see [68] for a first implementation). Early IR systems were mainly used by expert librarians as reference retrieval systems in batch modalities; indeed, many libraries still use categorization hierarchies to classify their volumes.

However, modern computers and the birth of the World Wide Web (1989) marked a permanent change to the concepts of storage, access, and searching of document collections, making them available to the general public and *indexing* them for precise and large-coverage retrieval.

As an academic discipline, IR has been defined in various ways [26]. Sections 1.1.1 and 1.1.2 discuss two definitions highlighting different interesting aspects that characterize IR: relevance and large, unstructured data sources.

S. Ceri et al., *Web Information Retrieval*, Data-Centric Systems and Applications,
DOI 10.1007/978-3-642-39314-3_1, © Springer-Verlag Berlin Heidelberg 2013

1.1.1 Defining Relevance

In [149], IR is defined as the discipline finding *relevant documents* as opposed to simple matches to lexical patterns in a query. This underlines a fundamental aspect of IR, i.e., that *the relevance of results is assessed relative to the information need, not the query*. Let us exemplify this by considering the information need of figuring out whether eating chocolate is beneficial in reducing blood pressure. We might express this via the search engine query: "chocolate effect pressure"; however, we will evaluate a resulting document as relevant if it addresses the information need, not just because it contains all the words in the query—although this would be considered to be a good relevance indicator by many IR models, as we will see later.

It may be noted that relevance is a concept with interesting properties. First, it is *subjective*: two users may have the same information need and give different judgments about the same retrieved document. Another aspect is its *dynamic* nature, both in space and in time: documents retrieved and displayed to the user at a given time may influence relevance judgments on the documents that will be displayed later. Moreover, according to his/her current status, a user may express different judgments about the same document (given the same query). Finally, relevance is *multifaceted*, as it is determined not just by the content of a retrieved result but also by aspects such as the authoritativeness, credibility, specificity, exhaustiveness, recency, and clarity of its source.

Note that relevance is not known to the system prior to the user's judgment. Indeed, we could say that the task of an IR system is to "guess" a set of documents D relevant with respect to a given query, say, q_k, by computing a relevance function $R(q_k, d_j)$ for each document $d_j \in D$. In Chap. 3, we will see that R depends on the adopted retrieval model.

1.1.2 Dealing with Large, Unstructured Data Collections

In [241], the IR task is defined as the task of finding documents characterized by an unstructured nature (usually text) that satisfy an information need from large collections, stored on computers.

A key aspect highlighted by this definition is the presence of *large collections*: our "digital society" has produced a large number of devices for the cost-free generation, storage, and processing of digital content. Indeed, while around 10^{18} bytes (10K petabytes) of information were created or replicated worldwide in 2006, 2010 saw this number increase by a factor of 6 (988 exabytes, i.e., nearly one zettabyte). These numbers correspond to about 10^6–10^9 documents, which roughly speaking exceeds the amount of written content created by mankind in the previous 5,000 years.

Finally, a key aspect of IR as opposed to, e.g., data retrieval is its *unstructured* nature. Data retrieval, as performed by relational database management systems

(RDBMSs) or Extensible Markup Language (XML) databases, refers to retrieving all objects that satisfy clearly defined conditions expressed through a formal query language. In such a context, data has a well-defined structure and is accessed via query languages with formal semantics, such as regular expressions, SQL statements, relational algebra expressions, etc. Furthermore, results are *exact* matches, hence partially correct matches are not returned as part of the response. Therefore, the ranking of results with respect to their relevance to the user's information need does not apply to data retrieval.

1.1.3 Formal Characterization

An *information retrieval model* (*IRM*) can be defined as a quadruple:

$$IRM = \{D, Q, F, R(q_k, d_j)\}$$

where

- D is a set of logical views (or representations) of the documents in the collection (referred to as d_j);
- Q is a set of logical views (or representations) of the user's information needs, called queries (referred to as q_k);
- F is a framework (or strategy) for modeling the representation of documents, queries, and their relationships;
- $R(q_k, d_j)$ is a ranking function that associates a real number to a document representation d_j, denoting its relevance to a query q_k.

The ranking function $R(q_k, d_j)$ defines a relevance order over the documents with respect to q_k and is a key element of the IR model. As illustrated in Chap. 3, different IR models can be defined according to R and to different query and document representations.

1.1.4 Typical Information Retrieval Tasks

Search engines are the most important and widespread application of IR, but IR techniques are also fundamental to a number of other tasks.

Information filtering systems remove redundant or undesired information from an information stream using (semi)automatic methods before presenting them to human users. Filtering systems typically compare a user's profile with a set of reference characteristics, which may be drawn either from information items (content-based approach) or from the user's social environment (collaborative filtering approach). A classic application of information filtering is that of spam filters, which learn to distinguish between useful and harmful emails based on the intrinsic content of the emails and on the users' behavior when processing them. The interested reader can refer to [153] for an overview of information filtering systems.

Document summarization is another IR application that consists in creating a shortened version of a text in order to reduce the information overload. Summarization is generally extractive; i.e., it proceeds by selecting the most relevant sentences from a document and collecting them to form a reduced version of the document itself. Reference [266] provides a contemporary overview of different summarization approaches and systems.

Document clustering and *categorization* are also important applications of IR. Clustering consists in grouping documents together based on their proximity (as defined by a suitable spatial model) in an unsupervised fashion. However, categorization starts from a predefined taxonomy of classes and assigns each document to the most relevant class. Typical applications of text categorization are the identification of news article categories or language, while clustering is often applied to group together dynamically created search results by their topical similarity. Chapter 4 provides an overview of document clustering and classification.

Question answering (QA) systems deal with the selection of relevant document portions to answer user's queries formulated in natural language. In addition to their capability of also retrieving answers to questions never seen before, the main feature of QA systems is the use of fine-grained relevance models, which provide answers in the form of relevant sentences, phrases, or even words, depending on the type of question asked (see Sect. 5.3). Chapter 5 illustrates the main aspects of QA systems.

Recommending systems may be seen as a form of information filtering, by which interesting information items (e.g., songs, movies, or books) are presented to users based on their profile or their neighbors' taste, neighborhood being defined by such aspects as geographical proximity, social acquaintance, or common interests. Chapter 8 provides an overview of this IR application.

Finally, an interesting aspect of IR concerns *cross-language retrieval*, i.e., the retrieval of documents formulated in a language different from the language of the user's query (see [270]). A notable application of this technology refers to the retrieval of legal documents (see, e.g., [313]).

1.2   Evaluating an Information Retrieval System

In Sect. 1.1.1, we have defined relevance as the key criterion determining IR quality, highlighting the fact that it refers to an *implicit* user need. How can we then identify the measurable properties of an IR system driven by subjective, dynamic, and multifaceted criteria? The remainder of this section answers these questions by outlining the desiderata of IR evaluation and discussing how they are met by adopting precision and recall as measurable properties.

1.2.1 Aspects of Information Retrieval Evaluation

The evaluation of IR systems should account for a number of desirable properties. To begin with, speed and efficiency of document processing would be useful evaluation criteria, e.g., by using as factors the number of documents retrieved per hour and their average size. Search speed would also be interesting, measured for instance by computing the latency of the IR system as a function of the document collection size and of the complexity and expressiveness of the query.

However, producing fast but useless answers would not make a user happy, and it can be argued that the ultimate objective of IR should be *user satisfaction*. Thus two vital questions to be addressed are: Who is the user we are trying to make happy? What is her behavior?

Providing an answer to the latter question depends on the application context. For instance, a satisfied Web search engine user will tend to return to the engine; hence, the rate of returning users can be part of the satisfaction metrics. On an e-commerce website, a satisfied user will tend to make a purchase: possible measures of satisfaction are the time taken to purchase an item, or the fraction of searchers who become buyers. In a company setting, employee "productivity" is affected by the time saved by employees when looking for information.

To formalize these issues, all of which refer to different aspects of relevance, we say that an IR system will be measurable in terms of relevance once the following information is available:

1. a benchmark collection D of documents,
2. a benchmark set Q of queries,
3. a tuple $t_{jk} = \langle d_j, q_k, r^* \rangle$ for each query $q_k \in Q$ and document $d_j \in D$ containing a binary judgment r^* of the relevance of d_j with respect to q_k, as assessed by a reference authority.

Section 1.2.2 illustrates the precision and recall evaluation metrics that usually concur to estimate the true value of r based on a set of documents and queries.

1.2.2 Precision, Recall, and Their Trade-Offs

When IR systems return unordered results, they can be evaluated appropriately in terms of *precision* and *recall*.

Loosely speaking, precision (P) is the fraction of retrieved documents that are relevant to a query and provides a measure of the "soundness" of the system.

Precision is not concerned with the total number of documents that are deemed relevant by the IR system. This aspect is accounted for by recall (R), which is defined as the fraction of "truly" relevant documents that are effectively retrieved and thus provides a measure of the "completeness" of the system.

Formally, given the complete set of documents D and a query q, let us define as $TP \subseteq D$ the set of true positive results, i.e., retrieved documents that are truly relevant to q. We define as $FP \subseteq D$ the set of false positives, i.e., the set of retrieved documents that are not relevant to q, and as $FN \subseteq D$ the set of documents that do correspond to the user's need but are not retrieved by the IR system. Given the above definitions, we can write

$$P = \frac{|TP|}{|TP| + |FP|}$$

and

$$R = \frac{|TP|}{|TP| + |FN|}$$

Computing TP, FP, and FN with respect to a document collection D and a set of queries Q requires obtaining reference assessments, i.e., the above-mentioned r^* judgment for each $q_k \in Q$ and $d_j \in D$. These should ideally be formulated by human assessors having the same background and a sufficient level of annotation agreement. Note that different domains may imply different levels of difficulty in assessing the relevance. Relevance granularity could also be questioned, as two documents may respond to the same query in correct but not equally satisfactory ways. Indeed, the precision and recall metrics suppose that the relevance of one document is assessed independently of any other document in the same collection.

As precision and recall have different advantages and disadvantages, a single balanced IR evaluation measure has been introduced as a way to mediate between the two components. This is called the *F-measure* and is defined as

$$F_\beta = \frac{(1 + \beta^2) \times P \times R}{(\beta^2 \times P) + R}$$

The most widely used value for β is 1, in order to give equal weight to precision and recall; the resulting measurement, the F_1-measure, is the harmonic mean of precision and recall.

Precision and recall normally are competing objectives. To obtain more relevant documents, a system lowers its selectivity and thus produces more false positives, with a loss of precision. To show the combined effect of precision and recall on the performance of an IR system, the *precision/recall plot* reports precision taken at different levels of recall (this is referred to as interpolated precision at a fixed recall level).

Recall levels are generally defined stepwise from 0 to 1, with 11 equal steps; hence, the interpolated precision P_{int} at a given level of recall R is measured as a function of the maximum subsequent level of recall R':

$$P_{\text{int}}(R) = \max_{R' \geq R} P(R')$$

As illustrated in Fig. 1, the typical precision/recall plot is monotonically decreasing: indeed, the larger the set of documents included in the evaluation, the more likely it becomes to include nonrelevant results in the final result set.

Fig. 1 A precision/recall plot. Precision is evaluated at 11 levels of recall

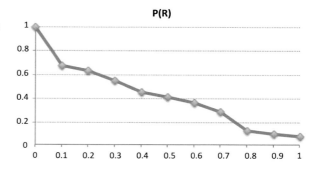

1.2.3 Ranked Retrieval

A notable feature of precision, recall, and their harmonic mean F_1 is that they do not take into account the rank of returned results, because true positives, false positives, and false negatives are treated as unordered sets for relevance computation.

In the context of ranked retrieval, when results are sorted by relevance and only a fraction of the retrieved documents are presented to the user, it is important to accurately select candidate results in order to maximize precision. An effective way to take into account the order by which documents appear in the result sets of a given query is to compute the gain in precision when augmenting the recall.

Average precision computes the average precision value obtained for the set of top k documents belonging to the result list after each *relevant* document is retrieved. The *average precision* of a query approximates the area of the (uninterpolated) precision/recall curve introduced in the previous section, and it is often computed as

$$AveP = \sum_{k=1}^{n} P(k)\Delta r(k)$$

where n is the number of retrieved documents for the query, $P(k)$ is the precision calculated when the result set is cut off at the relevant document k, and $\Delta r(k)$ is the variation in recall when moving from relevant document $k - 1$ to relevant document k.

Clearly, a precision measurement cannot be made on the grounds of the results for a single query. The precision of an IR engine is typically evaluated on the grounds of a set of queries representing its general usage. Such queries are often delivered together with standard test collections (see Sect. 1.2.4). Given the IR engine's results for a collection of Q queries, the *mean average precision* can then be computed as

$$MAP = \frac{\sum_{q=1}^{Q} AveP(q)}{Q}$$

Many applications, such as Web search, need to particularly focus on how many good results there are on the first page or the first few pages. A suitable evaluation metric would therefore be to measure the precision at a fixed—typically low—number of retrieved results, generally the first 10 or 30 documents. This measurement, referred to as *precision at k* and often abridged to $P@k$, has the advantage of not requiring any estimate of the size of the set of relevant documents, as the measure is evaluated after the first k documents in the result set. On the other hand, it is the least stable of the commonly used evaluation measures, and it does not average well, since it is strongly influenced by the total number of relevant documents for a query.

An increasingly adopted metric for ranked document relevance is *discounted cumulative gain (DCG)*. Like $P@k$, DCG is evaluated over the first k top search results. Unlike the previous metrics, which always assume a binary judgment for the relevance of a document to a query, DCG supports the use of a graded relevance scale.

DCG models the usefulness (gain) of a document based on its position in the result list. Such a gain is accumulated from the top of the result list to the bottom, following the assumptions that highly relevant documents are more useful when appearing earlier in the result list, and hence highly relevant documents appearing lower in a search result list should be penalized. The graded relevance value is therefore reduced logarithmically proportional to the position of the result in order to provide a smooth reduction rate, as follows:

$$DCG = rel_1 + \sum_{i=2}^{p} \frac{rel_i}{\log_2(i)}$$

where rel_1 is the graded relevance of the result at position i.

1.2.4 Standard Test Collections

Adopting effective relevance metrics is just one side of the evaluation: another fundamental aspect is the availability of reference document and query collections for which a relevance assessment has been formulated.

To account for this, document collections started circulating as early as the 1960s in order to enable head-to-head system comparison in the IR community. One of these was the *Cranfield collection* [91], consisting of 1,398 abstracts of aerodynamics journal articles, a set of 225 queries, and an exhaustive set of relevance judgments.

In the 1990s, the US National Institute of Standards and Technology (NIST) collected large IR benchmarks within the *TREC* Ad Hoc retrieval campaigns (trec.nist.gov). Altogether, this resulted in a test collection made of 1.89 million documents, mainly consisting of newswire articles; these are complete with relevance judgments for 450 "retrieval tasks" specified as queries compiled by human experts.

Since 2000, Reuters has made available a widely adopted resource for text classification, the *Reuters Corpus* Volume 1, consisting of 810,000 English-language news stories.[1] More recently, a second volume has appeared containing news stories in 13 languages (Dutch, French, German, Chinese, Japanese, Russian, Portuguese, Spanish, Latin American Spanish, Italian, Danish, Norwegian, and Swedish). To facilitate research on massive data collections such as blogs, the Thomson Reuters Text Research Collection (TRC2) has more recently appeared, featuring over 1.8 million news stories.[2]

Cross-language evaluation tasks have been carried out within the *Conference and Labs of the Evaluation Forum* (CLEF, www.clef-campaign.org), mostly dealing with European languages. The reference for East Asian languages and cross-language retrieval is the NII Test Collection for IR Systems *(NTCIR)*, launched by the Japan Society for Promotion of Science.[3]

1.3 Exercises

1.1 Given your experience with today's search engines, explain which typical tasks of information retrieval are currently provided in addition to ranked retrieval.

1.2 Compute the mean average precision for the precision/recall plot in Fig.1, knowing that it was generated using the following data:

R	P
0	1
0.1	0.67
0.2	0.63
0.3	0.55
0.4	0.45
0.5	0.41
0.6	0.36
0.7	0.29
0.8	0.13
0.9	0.1
1	0.08

1.3 Why is benchmarking against standard collections so important in evaluating information retrieval?

1.4 In what situations would you recommend aiming at maximum precision at the price of potentially lower recall? When instead would high recall be more important than high precision?

---

[1] See trec.nist.gov/data/reuters/reuters.html.

[2] Also at trec.nist.gov/data/reuters/reuters.html.

[3] research.nii.ac.jp/ntcir.

Chapter 2
The Information Retrieval Process

Abstract What does an information retrieval system look like from a bird's eye perspective? How can a set of documents be processed by a system to make sense out of their content and find answers to user queries? In this chapter, we will start answering these questions by providing an overview of the information retrieval process. As the search for text is the most widespread information retrieval application, we devote particular emphasis to textual retrieval. The fundamental phases of document processing are illustrated along with the principles and data structures supporting indexing.

2.1 A Bird's Eye View

If we consider the information retrieval (IR) process from a perspective of 10,000 feet, we might illustrate it as in Fig. 2.1.

Here, the user issues a query q from the front-end application (accessible via, e.g., a Web browser); q is processed by a *query interaction* module that transforms it into a "machine-readable" query q' to be fed into the core of the system, a *search and query analysis* module. This is the part of the IR system having access to the *content management* module directly linked with the back-end information source (e.g., a database). Once a set of results r is made ready by the search module, it is returned to the user via the result interaction module; optionally, the result is modified (into r') or updated until the user is completely satisfied.

The most widespread applications of IR are the ones dealing with textual data. As textual IR deals with document sources and questions, both expressed in natural language, a number of textual operations take place "on top" of the classic retrieval steps. Figure 2.2 sketches the processing of textual queries typically performed by an IR engine:

1. The user need is specified via the user interface, in the form of a textual *query* q_U (typically made of keywords).
2. The query q_U is parsed and transformed by a set of textual operations; the same operations have been previously applied to the contents indexed by the IR system (see Sect. 2.2); this step yields a refined query q_U'.
3. Query operations further transform the preprocessed query into a system-level representation, q_S.

Fig. 2.1 A high-level view of the IR process

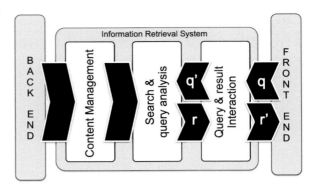

Fig. 2.2 Architecture of a textual IR system. Textual operations translate the user's need into a logical query and create a logical view of documents

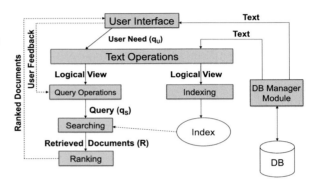

4. The query q_S is executed on top of a document source D (e.g., a text database) to retrieve a set of relevant documents, R. Fast query processing is made possible by the index structure previously built from the documents in the document source.
5. The set of retrieved documents R is then ordered: documents are ranked according to the estimated relevance with respect to the user's need.
6. The user then examines the set of ranked documents for useful information; he might pinpoint a subset of the documents as definitely of interest and thus provide feedback to the system.

Textual IR exploits a sequence of text operations that translate the user's need and the original content of textual documents into a logical representation more amenable to indexing and querying. Such a "logical", machine-readable representation of documents is discussed in the following section.

2.1.1 Logical View of Documents

It is evident that on-the-fly scanning of the documents in a collection each time a query is issued is an impractical, often impossible solution. Very early in the history of IR it was found that avoiding linear scanning requires *indexing* the documents in advance.

The index is a logical view where documents in a collection are represented through a set of *index terms* or *keywords*, i.e., any word that appears in the document text. The assumption behind indexing is that the semantics of both the documents and the user's need can be properly expressed through sets of index terms; of course, this may be seen as a considerable oversimplification of the problem. Keywords are either extracted directly from the text of the document or specified by a human subject (e.g., tags and comments). Some retrieval systems represent a document by the complete set of words appearing in it (logical full-text representation); however, with very large collections, the set of representative terms has to be reduced by means of *text operations*. Section 2.2 illustrates how such operations work.

2.1.2   Indexing Process

The indexing process consists of three basic steps: defining the data source, transforming document content to generate a logical view, and building an index of the text on the logical view.

In particular, data source definition is usually done by a database manager module (see Fig. 2.2), which specifies the documents, the operations to be performed on them, the content structure, and what elements of a document can be retrieved (e.g., the full text, the title, the authors). Subsequently, the text operations transform the original documents and generate their logical view; an index of the text is finally built on the logical view to allow for fast searching over large volumes of data. Different index structures might be used, but the most popular one is the inverted file, illustrated in Sect. 2.3.

2.2   A Closer Look at Text

When we consider natural language text, it is easy to notice that not all words are equally effective for the representation of a document's semantics. Usually, noun words (or word groups containing nouns, also called noun phrase groups) are the most representative components of a document in terms of content. This is the implicit mental process we perform when distilling the "important" query concepts into some representative nouns in our search engine queries. Based on this observation, the IR system also preprocesses the text of the documents to determine the most "important" terms to be used as index terms; a subset of the words is therefore selected to represent the content of a document.

When selecting candidate keywords, indexing must fulfill two different and potentially opposite goals: one is *exhaustiveness*, i.e., assigning a sufficiently large number of terms to a document, and the other is *specificity*, i.e., the exclusion of

Fig. 2.3 Text processing
phases in an IR system

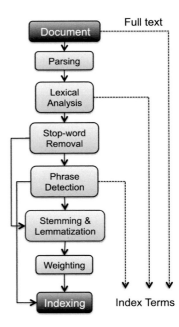

generic terms that carry little semantics and inflate the index. Generic terms, for example, conjunctions and prepositions, are characterized by a low discriminative power, as their frequency across any document in the collection tends to be high. In other words, generic terms have high *term frequency*, defined as the number of occurrences of the term in a document. In contrast, specific terms have higher discriminative power, due to their rare occurrences across collection documents: they have low *document frequency*, defined as the number of documents in a collection in which a term occurs.

2.2.1 Textual Operations

Figure 2.3 sketches the textual preprocessing phase typically performed by an IR engine, taking as input a document and yielding its index terms.

1. *Document Parsing.* Documents come in all sorts of languages, character sets, and formats; often, the same document may contain multiple languages or formats, e.g., a French email with Portuguese PDF attachments. *Document parsing* deals with the recognition and "breaking down" of the document structure into individual components. In this preprocessing phase, unit documents are created; e.g., emails with attachments are split into one document representing the email and as many documents as there are attachments.
2. *Lexical Analysis.* After parsing, *lexical analysis* tokenizes a document, seen as an input stream, into words. Issues related to lexical analysis include the correct

identification of accents, abbreviations, dates, and cases. The difficulty of this operation depends much on the language at hand: for example, the English language has neither diacritics nor cases, French has diacritics but no cases, German has both diacritics and cases. The recognition of abbreviations and, in particular, of time expressions would deserve a separate chapter due to its complexity and the extensive literature in the field; the interested reader may refer to [18, 227, 239] for current approaches.

3. *Stop-Word Removal.* A subsequent step optionally applied to the results of lexical analysis is *stop-word removal*, i.e., the removal of high-frequency words. For example, given the sentence *"search engines are the most visible information retrieval applications"* and a classic stop words set such as the one adopted by the Snowball stemmer,[1] the effect of stop-word removal would be: *"search engine most visible information retrieval applications"*.

    However, as this process may decrease recall (prepositions are important to disambiguate queries), most search engines do not remove them [241]. The subsequent phases take the full-text structure derived from the initial phases of parsing and lexical analysis and process it in order to identify relevant keywords to serve as index terms.

4. *Phrase Detection.* This step captures text meaning beyond what is possible with pure bag-of-word approaches, thanks to the identification of noun groups and other phrases. Phrase detection may be approached in several ways, including rules (e.g., retaining terms that are not separated by punctuation marks), morphological analysis (part-of-speech tagging—see Chap. 5), syntactic analysis, and combinations thereof. For example, scanning our example sentence *"search engines are the most visible information retrieval applications"* for noun phrases would probably result in identifying *"search engines"* and *"information retrieval"*.

    A common approach to phrase detection relies on the use of thesauri, i.e., classification schemes containing words and phrases recurrent in the expression of ideas in written text. Thesauri usually contain synonyms and antonyms (see, e.g., *Roget's Thesaurus* [297]) and may be composed following different approaches. Human-made thesauri are generally hierarchical, containing related terms, usage examples, and special cases; other formats are the associative one, where graphs are derived from document collections in which edges represent semantic associations, and the clustered format, such as the one underlying WordNet's synonym sets or *synsets* [254].

    An alternative to the consultation of thesauri for phrase detection is the use of machine learning techniques. For instance, the Key Extraction Algorithm (KEA) [353] identifies candidate keyphrases using lexical methods, calculates feature values for each candidate, and uses a supervised machine learning algorithm to predict which candidates are good phrases based on a corpus of previously annotated documents.

---

[1] http://snowball.tartarus.org/algorithms/english/stop.txt.

5. *Stemming and Lemmatization.* Following phrase extraction, *stemming* and *lemmatization* aim at stripping down word suffixes in order to normalize the word. In particular, stemming is a heuristic process that "chops off" the ends of words in the hope of achieving the goal correctly most of the time; a classic rule-based algorithm for this was devised by Porter [280]. According to the Porter stemmer, our example sentence *"Search engines are the most visible information retrieval applications"* would result in: *"Search engin are the most visibl inform retriev applic"*.

Lemmatization is a process that typically uses dictionaries and morphological analysis of words in order to return the base or dictionary form of a word, thereby collapsing its inflectional forms (see, e.g., [278]). For example, our sentence would result in *"Search engine are the most visible information retrieval application"* when lemmatized according to a WordNet-based lemmatizer.[2]

6. *Weighting.* The final phase of text preprocessing deals with *term weighting*. As previously mentioned, words in a text have different descriptive power; hence, index terms can be weighted differently to account for their significance within a document and/or a document collection. Such a weighting can be binary, e.g., assigning 0 for term absence and 1 for presence. Chapter 3 illustrates different IR models exploiting different weighting schemes to index terms.

2.2.2 Empirical Laws About Text

Some interesting properties of language and its usage were studied well before current IR research and may be useful in understanding the indexing process.

1. *Zipf's Law.* Formulated in the 1940s, Zipf's law [373] states that, given a corpus of natural language utterances, the frequency of any word is inversely proportional to its rank in the frequency table. This can be empirically validated by plotting the frequency of words in large textual corpora, as done for instance in a well-known experiment with the Brown Corpus.[3] Formally, if the words in a document collection are ordered according to a ranking function $r(w)$ in decreasing order of frequency $f(w)$, the following holds:

$$r(w) \times f(w) = c$$

where c is a language-dependent constant. For instance, in English collections c can be approximated to 10.

2. *Luhn's Analysis.* Information from Zipf's law can be combined with the findings of Luhn, roughly ten years later: "It is here proposed that the frequency of word

---

[2]See http://text-processing.com/demo/stem/.

[3]The Brown Corpus was compiled in the 1960s by Henry Kucera and W. Nelson Francis at Brown University, Providence, RI, as a general corpus containing 500 samples of English-language text, involving roughly one million words, compiled from works published in the United States in 1961.

occurrence in an article furnishes a useful measurement of word significance. It is further proposed that the relative position within a sentence of words having given values of significance furnish a useful measurement for determining the significance of sentences. The significance factor of a sentence will therefore be based on a combination of these two measurements." [233].

Formally, let $f(w)$ be the frequency of occurrence of various word types in a given position of text and $r(w)$ their rank order, that is, the order of their frequency of occurrence; a plot relating $f(w)$ and $r(w)$ yields a hyperbolic curve, demonstrating Zipf's assertion that the product of the frequency of use of words and the rank order is approximately constant.

Luhn used this law as a null hypothesis to specify two cut-offs, an upper and a lower, to exclude nonsignificant words. Indeed, words above the upper cut-off can be considered as too common, while those below the lower cut-off are too rare to be significant for understanding document content. Consequently, Luhn assumed that the *resolving power* of significant words, by which he meant the ability of words to discriminate content, reached a peak at a rank order position halfway between the two cut-offs and from the peak fell off in either direction, reducing to almost zero at the cut-off points.

3. *Heap's Law.* The above findings relate the frequency and relevance of words in a corpus. However, an interesting question regards how *vocabulary* grows with respect to the *size* of a document collection. Heap's law [159] has an answer for this, stating that the vocabulary size V can be computed as

$$V = K N^\beta$$

where N is the size (in words) of the document collection, K is a constant (typically between 10 and 100), and $0 < \beta < 1$ is a constant, typically between 0.4 and 0.6.

This finding is very important for the scalability of the indexing process: it states that vocabulary size (and therefore index size) exhibits a less-than-linear growth with respect to the growth of the document collection. In a representation such as the vector space model (see Sect. 3.3), this means that the dimension of the vector space needed to represent very large data collections is not necessarily much higher than that required for a small collection.

2.3 Data Structures for Indexing

Let us now return to the indexing process of translating a document into a set of relevant terms or keywords. The first step requires defining the text data source. This is usually done by the database manager (see Fig. 2.2), who specifies the documents, the operations to be performed on them, and the content model (i.e., the content structure and what elements can be retrieved). Then, a series of content operations transform each original document into its logical representation; an index of the text is built on such a logical representation to allow for fast searching over large

Table 2.1 An inverted index: each word in the dictionary (i.e., posting) points to a list of documents containing the word (posting list)

Dictionary entry	Posting list for entry
⋮	…
princess	1 3 8 22 41 55 67 68 78 120
⋮	…
witch	1 2 8 30 …
⋮	…
dragon	2 3 4 122 …
⋮	…

volumes of data. The rationale for indexing is that the cost (in terms of time and storage space) spent on building the index is progressively repaid by querying the retrieval system many times.

The first question to address when preparing indexing is therefore what storage structure to use in order to maximize retrieval efficiency. A naive solution would just adopt a *term-document incidence matrix*, i.e., a matrix where rows correspond to terms and columns correspond to documents in a collection C, such that each cell c_{ij} is equal to 1 if term t_i occurs in document d_j, and 0 otherwise. However, in the case of large document collections, this criterion would result in a very sparse matrix, as the probability of each word to occur in a collection document decreases with the number of documents. An improvement over this situation is the *inverted index*, described in Sect. 2.3.1.

2.3.1 Inverted Indexes

The principle behind the inverted index is very simple. First, a *dictionary* of terms (also called a vocabulary or lexicon), V, is created to represent all the unique occurrences of terms in the document collection C. Optionally, the frequency of appearance of each term $t_i \in V$ in C is also stored. Then, for each term $t_i \in V$, called the *posting*, a list L_i, the *posting list* or *inverted list*, is created containing a reference to each document $d_j \in C$ where t_i occurs (see Table 2.1). In addition, L_i may contain the frequency and position of t_i within d_j. The set of postings together with their posting lists is called the *inverted index* or *inverted file* or *postings file*.

Let us assume we intend to create an index for a corpus of fairy tales, sentences from which are reported in Table 2.2 along with their documents.

First, a mapping is created from each word to its document (Fig. 2.4(a)); the subdivision in sentences is no longer considered. Subsequently, words are sorted (alphabetically, in the case of Fig. 2.4(b)); then, multiple occurrences are merged and their total frequency is computed—document wise (Fig. 2.4(c)). Finally, a dictionary is created together with posting lists (Fig. 2.5); the result is the inverted index of Fig. 2.1.

Word	Document ID
once	1
upon	1
a	1
time	1
there	1
lived	1
a	1
beautiful	1
princess	1
the	1
witch	1
cast	1
a	1
terrible	1
spell	1
on	1
the	1
princess	1
the	2
witch	2
hunted	2
the	2
dragon	2
down	2
tThe	2
dragon	2
fought	2
back	2
but	2
the	2
witch	2
was	2
stronger	2

(a) map

Word	Document ID
a	1
a	1
a	1
back	2
beautiful	1
but	2
cast	1
down	2
dragon	2
dragon	2
fought	2
hunted	2
lived	1
on	1
once	1
princess	1
princess	1
spell	1
stronger	2
terrible	1
the	1
the	1
the	2
the	2
the	2
the	2
time	1
there	1
upon	1
was	2
witch	1
witch	2
witch	2

(b) sort

Word	Document ID	Frequency
a	1	3
back	2	1
beautiful	1	1
but	2	1
cast	1	1
down	2	1
dragon	2	2
fought	2	1
hunted	2	1
lived	1	1
on	1	1
once	1	1
princess	1	2
spell	1	1
stronger	2	1
terrible	1	1
the	1	2
the	2	4
time	1	1
there	1	1
upon	1	1
was	2	1
witch	1	1
witch	2	2

(c) merge

Fig. 2.4 Index creation. (**a**) A mapping is created from each sentence word to its document, (**b**) words are sorted, (**c**) multiple word entries are merged and frequency information is added

Inverted indexes are unrivaled in terms of retrieval efficiency: indeed, as the same term generally occurs in a number of documents, they reduce the storage requirements. In order to further support efficiency, linked lists are generally preferred to arrays to represent posting lists, despite the space overhead of pointers, due to their dynamic space allocation and the ease of term insertion.

2.3.2 Dictionary Compression

The Heap law (Sect. 2.2.2(3)) tells us that the growth of a dictionary with respect to vocabulary size is $O(n^{\beta})$, with $0.4 < \beta < 0.6$; this means that the size of the vocabulary represented in a 1 Gb document set would roughly fit in about 5 Mb,

Table 2.2 Example documents from a fairy tale corpus

Document ID	sentence ID	text
1	1	Once upon a time there lived a beautiful princess
		⋮
1	19	The witch cast a terrible spell on the princess
2	34	The witch hunted the dragon down
		⋮
2	39	The dragon fought back but the witch was stronger

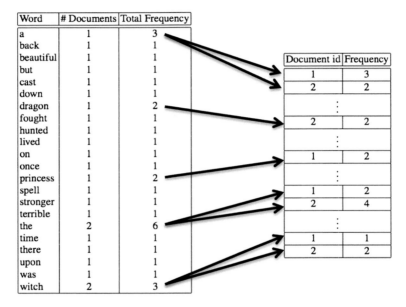

Fig. 2.5 Index creation: a dictionary is created together with posting lists

i.e., a reasonably sized file. In other words, the size of a dictionary representing a document collection is generally sufficiently small to be stored in memory. In contrast, posting lists are generally kept on disk as they are directly proportional to the number of documents; i.e., they are $O(n)$.

However, in the case of very large data collections, dictionaries need to be compressed in order to fit into memory. Besides, while the advantages of a linear index (i.e., one where the vocabulary is a sequential list of words) include low *access time* (e.g., $O(\log(n))$ in the case of binary search) and low space occupation, their construction is an elaborate process that occurs at each insertion of a new document.

To counterbalance such issues, efficient dictionary storage techniques have been devised, including string storage and block storage.

- In *string storage*, the index may be represented either as an array of fixed-width entries or as long strings of characters coupled with pointers for locating terms in such strings. This way, dictionary size can be reduced to as far as one-half of the space required for the array representation.
- In *block storage*, string terms are grouped into blocks of fixed size k and a pointer is kept to the first term of each block; the length of the term is stored as an additional byte. This solution eliminates $k - 1$ term pointers but requires k additional bytes for storing the length of each term; the choice of a block size is a trade-off between better compression and slower performance.

2.3.3 B and B+ Trees

Given the data structures described above, the process of searching in an inverted index structure consists of four main steps:

1. First, the dictionary file is accessed to identify query terms;
2. then, the posting files are retrieved for each query term;
3. then, results are filtered: if the query is composed of several terms (possibly connected by logical operators), partial result lists must be fused together;
4. finally, the result list is returned.

As searching arrays is not the most efficient strategy, a clever alternative consists in the representation of *indexes as search trees*. Two alternative approaches employ *B-trees* and their variant *B+ trees*, both of which are generalizations of binary search trees to the case of nodes with more than two children. In B-trees (see Fig. 2.6), internal (non-leaf) nodes contain a number of keys, generally ranging from d to $2d$, where d is the tree depth. The number of branches starting from a node is 1 plus the number of keys in the node. Each key value K_i is associated with two pointers (see Fig. 2.7): one points directly to the block (subtree) that contains the entry corresponding to K_i (denoted $t(K_i)$), while the second one points to a subtree with keys greater than K_i and less than K_{i+1}.

Searching for a key K in a B-tree is analogous to the search procedure in a binary search tree. The only difference is that, at each step, the possible choices are not two but coincide with the number of children of each node. The recursive procedure starts at the B-tree root node. If K is found, the search stops. Otherwise, if K is smaller than the leftmost key in the node, the search proceeds following the node's leftmost pointer (p_0 in Fig. 2.7); if K is greater than the rightmost key in the node, the search proceeds following the rightmost pointer (p_F in Fig. 2.7); if K is comprised between two keys of the node, the search proceeds within the corresponding node (pointed to by p_i in Fig. 2.7).

The maintenance of B-trees requires two operations: insertion and deletion. When the insertion of a new key value cannot be done locally to a node because it is full (i.e., it has reached the maximum number of keys supported by the B-tree structure), the median key of the node is identified, two child nodes are created,

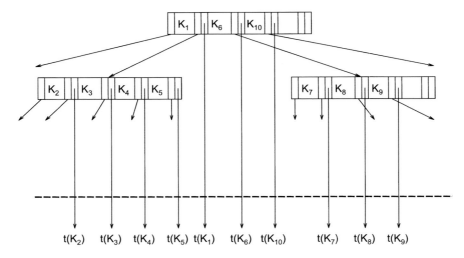

Fig. 2.6 A B-tree. The first key K_1 in the top node has a pointer to $t(K_1)$ and a pointer to a subtree containing all keys between K_1 and the following key in the top node, K_6: these are K_2, K_3, K_4, and K_5

Fig. 2.7 A B-tree node. Each key value K_i has two pointers: the first one points directly to the block that contains the entry corresponding to K_i, while the second points to a subtree with keys greater than K_i and less than $K_i + 1$

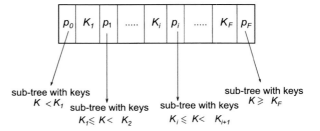

each containing the same number of keys, and the median key remains in the current node, as illustrated in Fig. 2.8 (insertion of K_3).

When a key is deleted, two "nearby" nodes have entries that could be condensed into a single node in order to maintain a high node filling rate and minimal paths from the root to the leaves: this is the merge procedure illustrated in Fig. 2.8 (deletion of K_2). As it causes a decrease of pointers in the upper node, one merge may recursively cause another merge.

A B-tree is kept balanced by requiring that all leaf nodes be at the same depth. This depth will increase slowly as elements are added to the tree, but an increase in the overall depth is infrequent, and results in all leaf nodes being one more node further away from the root.

Fig. 2.8 Insertion and
deletion in a B-tree

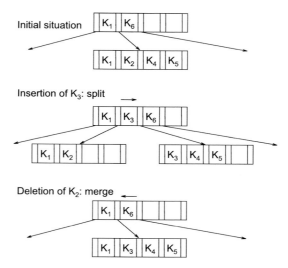

2.3.4 Evaluation of B and B+ Trees

B-trees are widely used in *relational database management systems* (RDBMSs) because of their short access time: indeed, the maximum number of accesses for a
B-tree of order d is $O(\log_d n)$, where n is the depth of the B-tree. Moreover, B-trees
are effective for updates and insertion of new terms, and they occupy little space.

However, a drawback of B-trees is their poor performance in sequential search.
This issue can be managed by the B+ tree variant, where leaf nodes are linked forming a chain following the order imposed by a key. Another disadvantage of B-trees
is that they may become unbalanced after too many insertions; this can be amended
by adopting rebalancing procedures.

Alternative structures to B-trees and B+ trees include suffix-tree arrays, where
document text is managed as a string, and each position in the text until the end is
a suffix (each suffix is uniquely indexed). The latter are typically used in genetic
databases or in applications involving complex search (e.g., search by phrases).
However, they are expensive to construct, and their index size is inevitably larger
than the document base size (generally by about 120–240 %).

2.4 Exercises

2.1 Apply the Porter stemmer[4] to the following quote from J.M. Barrie's *Peter Pan*:

> When a new baby laughs for the first time a new fairy is born, and as there
> are always new babies there are always new fairies.

---

[4]http://tartarus.org/~martin/PorterStemmer/.

Table 2.3 Collection of documents about information retrieval

Document	content
D_1	information retrieval students work hard
D_2	hard-working information retrieval students take many classes
D_3	the information retrieval workbook is well written
D_4	the probabilistic model is an information retrieval paradigm
D_5	the Boolean information retrieval model was the first to appear

How would a representation of the above sentence in terms of a bag of stems differ from a bag-of-words representation? What advantages and disadvantages would the former representation offer?

2.2 Draw the term-incidence matrix corresponding to the document collection in Table 2.3.

2.3 Recreate the inverted index procedure outlined in Sect. 2.3.1 using the document collection in Table 2.3.

2.4 Summarize the practical consequences of Zip's law, Luhn's analysis, and Heap's law.

2.5 Apply the six textual transformations outlined in Sect. 2.2.1 to the text in document D_2 from Table 2.3. Use a binary scheme and the five-document collection above as a reference for weighting.

Chapter 3
Information Retrieval Models

Abstract This chapter introduces three classic information retrieval models: Boolean, vector space, and probabilistic. These models provide the foundations of query evaluation, the process that retrieves the relevant documents from a document collection upon a user's query. The three models represent documents and compute their relevance to the user's query in very different ways. We illustrate each of them separately and then compare their features.

3.1 Similarity and Matching Strategies

So far, we have been discussing the representation of documents and queries and the techniques for document indexing. This is only part of the retrieval process. Another fundamental issue is the method for determining the degree of relevance of the user's query with respect to the document representation, also called the *matching process*. In most practical cases, this process is expected to produce a ranked list of documents, where relevant documents should appear towards the top of the ranked list, in order to minimize the time spent by users in identifying relevant information.

Ranking algorithms may use a variety of information sources, the frequency distribution of terms over documents, as well as other proprieties, e.g., in the Web search context, the "social relevance" of a page determined from the links that point to it.

In this chapter, we introduce three classic information retrieval (IR) models. We start with the Boolean model, described in Sect. 3.2, the first IR model and probably also the most basic one. It provides exact matching; i.e., documents are either retrieved or not, and thus supports the construction of result sets in which documents are not ranked.

Then, we follow Luhn's intuition of adopting a statistical approach for IR [232]: he suggested to use the degree of similarity between an artificial document constructed from the user's query and the representation of documents in the collection as a relevance measure for ranking search results. A simple way to do so is by counting the number of elements that are shared by the query and by the index representation of the document. This is the principle behind the vector space model, discussed in Sect. 3.3.

S. Ceri et al., *Web Information Retrieval*, Data-Centric Systems and Applications, DOI 10.1007/978-3-642-39314-3_3, © Springer-Verlag Berlin Heidelberg 2013

Last, we illustrate the probabilistic indexing model. Unlike the previous ones, this model was not meant to support automatic indexing by IR systems; rather, it assumed a human evaluator to manually provide a probability value for each index term to be relevant to a document. An adaptation of this idea suitable for automatic IR is discussed in Sect. 3.4.

Before we discuss the details of each specific model, let us first introduce a simple definition that we will use throughout the chapter. An IR model can be defined as an algorithm that takes a query q and a set of documents $D = \{d_1, \ldots, d_N\}$ and associates a *similarity coefficient* with respect to q, $SC(q, d_i)$ to each of the documents d_i, $1 \leq i \leq N$. The latter is also called the *retrieval status value*, and is abbreviated as *rsv*.

3.2 Boolean Model

The Boolean retrieval model [213] is a simple retrieval model based on set theory and Boolean algebra, whereby queries are defined as Boolean expressions over index terms (using the Boolean operators *AND*, *OR*, and *NOT*), e.g., "*photo AND mountain OR snow*".

The significance of an index term t_i is represented by binary weights: a weight $w_{ij} \in 0, 1$ is associated to the tuple (t_i, d_j) as a function of $R(d_j)$, the set of index terms in d_i, and $R(t_i)$, the set of documents where the index term appears.

Relevance with respect to a query is then modeled as a binary-valued property of each document (hence either $SC(q, d_j) = 0$ or $SC(q, d_j) = 1$), following the (strong) closed world assumption by which the absence of a term t in a document d is equivalent to the presence of $\neg t$ in the same representation.

3.2.1 Evaluating Boolean Similarity

A Boolean query q can be rewritten in disjunctive normal form (DNF), i.e., as a disjunction of conjunctive clauses. For instance,

$$q = t_a \wedge (t_b \vee \neg t_c)$$

can be rewritten as

$$q_{\text{DNF}} = (t_a \wedge t_b \wedge t_c) \vee (t_a \wedge t_b \wedge \neg t_c) \vee (t_a \wedge \neg t_b \wedge \neg t_c)$$

Note that the entire set of documents is partitioned into eight disjoint subsets, each represented by a given conjunctive clause. Then, the query is considered to be satisfied by a given document if this satisfies at least one of its conjunctive clauses. Given q above and its DNF representation q_{DNF}, we can consider the query to be satisfied for the following combinations of weights associated to terms t_a, t_b, and t_c:

$$q_{\text{DNF}} = (1, 1, 1) \vee (1, 1, 0) \vee (1, 0, 0)$$

Hence

$$SC(q, d_j) = \begin{cases} 1 & \exists \, q_c | (q_c \in q_{DNF}) \wedge (\forall \, t_i \in q_c \rightarrow w_{ij} = 1) \\ & \wedge \, (\forall \, t_i \notin q_c \rightarrow w_{ij} = 0) \\ 0 & \text{otherwise} \end{cases}$$

A Boolean query q can be computed by retrieving all the documents containing its terms and building a list for each term. Once such lists are available, Boolean operators must be handled as follows:

q_1 OR q_2 requires building the *union* of the lists of q_1 and q_2;
q_1 AND q_2 requires building the *intersection* of the lists of q_1 and q_2;
q_1 AND NOT q_2 requires building the *difference* of the lists of q_1 and q_2.

For example, computing the result set of the query $t_a \wedge t_b$ implies the five following steps:

1. locating t_a in the dictionary;
2. retrieving its postings list L_a;
3. locating t_b in the dictionary;
4. retrieving its postings list L_b;
5. intersecting L_a and L_b.

A standard heuristic for query optimization, i.e., for minimizing the total amount of work performed by the system, consists in processing terms in increasing order of term frequency, starting with small postings lists. However, optimization must also reflect the correct order of evaluation of logical expressions, i.e., give priority first to the *AND* operator, then to *OR*, then to *NOT*.

3.2.2 Extensions and Limitations of the Boolean Model

Extensions of the Boolean model allow for keyword truncation, i.e., using the wild-card character $*$ to signal the acceptance of partial term matches (*sax* OR viol**). Other extensions include the support for information adjacency and distance, as encoded by proximity operators. The latter are a way of specifying that two terms in a query must occur close to each other in a document, where closeness may be measured by limiting the allowed number of intervening words or by reference to a structural unit such as a sentence or paragraph (*rock NEAR roll*).

Being based on a binary decision criterion (i.e., a document is considered to be either relevant or nonrelevant), the Boolean model is in reality much more of a data retrieval model borrowed from the database realm than an IR model. As such, it shares some of the advantages as well as a number of limitations of the database approach. First, Boolean expressions have precise semantics, making them suitable for structured queries formulated by "expert" users. The latter are more likely to formulate *faceted queries*, involving the disjunction of quasi-synonyms (facets) joined via *AND*, for example, *(jazz OR classical) AND (sax OR clarinet OR flute) AND (Parker OR Coltrane)*.

The simple and neat formalism offered by the Boolean model has granted it great attention in the past years, and it was adopted by many early commercial bibliography retrieval systems. On the other hand, in many search contexts it is not simple to translate an information need into a Boolean expression. Moreover, Boolean expressions, returning exact result sets, do not support approximate matches or *ranking*, hence no indication of result quality is available. As illustrated in the remainder of this chapter, different models address this issue.

3.3 Vector Space Model

The vector space model (VSM) [301, 304] represents documents and queries as vectors in a space where each dimension corresponds to a term in their common vocabulary.

The idea underlying this model is the following. First, each term t_i in the dictionary V is represented by means of a *base* in a $|V|$-dimensional orthonormal space. In other words, each $\mathbf{t}_i$ contains all 0s but one 1 corresponding to the dimension associated with t_i in the space: for instance, in a 10-dimensional vector space where the term "jazz" is mapped to the third dimension, we will have $\mathbf{t}_{\text{jazz}} = \{0, 0, 1, 0, 0, 0, 0, 0, 0, 0\}$.

Any query q and document $d_j \in D$ may be represented in the vector space as:

$$\mathbf{q} = \sum_{i=1}^{|V|} w_{iq} \cdot \mathbf{t}_i, \qquad \mathbf{d}_j = \sum_{i=1}^{|V|} w_{ij} \cdot \mathbf{t}_i$$

where w_{iq} and w_{ij} are weights assigned to terms t_i for the query q and for each document d_j, according to a chosen weighting scheme. This model assumes that documents represented by "nearby" vectors in the vector space "talk about" the same things; hence, they will satisfy queries represented by close-by vectors in the same vector space. In other words, document vectors and query vectors can be used to compute a degree of similarity between documents or between documents and queries. Such a similarity can intuitively be represented as the projection of one vector over another, mathematically expressed in terms of the dot product; a query-document *similarity score* may therefore be computed as:

$$SC(d_j, q) = sim(d_j, q) = \mathbf{d}_j \cdot \mathbf{q}$$

3.3.1 Evaluating Vector Similarity

The advantage of modeling documents and queries in a vector space is, as previously mentioned, the ability to define a *similarity* metric between them, i.e., a measurement with the following properties:

- positive definiteness: $sim(d_1, d_2) \geq 0$, $\forall\, d_1, d_2$;
- symmetry: $sim(d_1, d_2) = sim(d_2, d_1)$;
- auto-similarity: $sim(d_1, d_1)$ is the maximum possible similarity.

The Euclidean dot product in the orthonormal vector space defined by the terms in $t_i \in V$ represents an example of such a metric, also called *cosine similarity*:

$$\mathbf{d}_j \cdot \mathbf{q} = \|\mathbf{d}_j\| \times \|\mathbf{q}\| \times \cos(\alpha)$$

Cosine similarity is a function of the size of the angle α formed by $\mathbf{d}_j$ and $\mathbf{q}$ in the space:

$$sim_{\cos}(d_j, q) = \cos(\alpha) = \frac{\mathbf{d}_j \cdot \mathbf{q}}{\|\mathbf{d}_j\| \times \|\mathbf{q}\|} = \frac{\sum_{i=1}^{|V|}(w_{ij} \times w_{iq})}{\sqrt{\sum_{i=1}^{|V|}(w_{ij})^2} \times \sqrt{\sum_{i=1}^{|V|}(w_{iq})^2}}$$

The cosine measure performs a normalization of vector length, e.g., by using the Euclidean or L_2-norm (represented as $\|\cdot\|$). This allows one to rule out both false negatives, when vectors of very different lengths result in negligible projections, and false positives, where "large" vectors match just because of their length.

Other similarity functions include Jaccard and Dice similarity and kernel functions. Jaccard similarity [185] is defined as the size of the intersection of two sets divided by the size of their union: hence, if representing the set of query terms as Q and the set of a document's terms as D,

$$sim_{\text{Jac}}(D, Q) = \frac{|D \cap Q|}{|D \cup Q|}$$

Analogously, the Dice similarity formula [108] is defined as

$$sim_{\text{Dice}}(D, Q) = \frac{2|D \cap Q|}{|D| + |Q|}$$

We refer to Chap. 5 for an account of kernel functions and their usage in an important IR application called question answering.

3.3.2 Weighting Schemes and $tf \times idf$

The choice of a similarity function is not the only factor affecting the evaluation of relevance in IR. The adopted weighting scheme is also vital to determine the position in space of each document vector. Indeed, vector normalization might not be sufficient to compensate for differences in document lengths; for instance, a longer document has more opportunity to have some components that are relevant to a query. More importantly, a binary weighting scheme does not distinguish between terms having different discriminative power.

These observations suggest adopting a more sophisticated weighting scheme than the binary scheme. One might for instance think of a weighting in which the more a document discusses a term, the more it will match a query containing such a term: given a query as simple as a single keyword, this criterion will make sure that a document with many occurrences of such a keyword will result as more relevant than another having only one. On the other hand, one might want to consider the distribution of terms not only in individual documents forming the collection D, but also across documents; for instance, the term "actor" should have a different weight in a collection of news articles than in an online movie database.

A well-known weighting scheme taking both these aspects into account is the $tf \times idf$ formula, which defines the weight of a term i in a document d_j within a collection D as:

$$w_{ij} = tf_{ij} \times idf_i$$

i.e., the frequency of i within d_j times the inverse of the frequency of the documents that contain i with respect to all the documents in the entire collection D. The weight of a term in a document therefore tends to increase with its frequency within the document itself and with its rarity within the document collection.

3.3.3 Evaluation of the Vector Space Model

So far, we have considered two retrieval models: the Boolean model and the VSM. Some advantages of the VSM with respect to the Boolean model reside in its flexible and intuitive geometric interpretation, the possibility of weighting the representations of both queries and documents, and the avoidance of output "flattening," as each query term contributes to the ranking based on its weight. This makes the results of an IR engine based on the VSM much more effective than those based on a Boolean model. However, this comes at the cost of making it impossible to formulate "exact" queries, because, unlike the Boolean model, the VSM does not support operators. In addition, the VSM requires an assumption of stochastic independence of terms (see Sect. 3.4.1), which is not always appropriate, although it works well in practice. Finally, the efficiency of IR engines based on the VSM is highly dependent on the size of the vector space. Even when preprocessing such as stemming is applied, the vector space resulting from a large document collection may easily exceed 10^5 dimensions.

3.4 Probabilistic Model

IR is an uncertain process: the criteria by which the information need is converted into a query, documents are converted into index terms, and query/document term matches are computed are far from exact. A probabilistic IR model [131] attempts to represent the probability of relevance of a document given a query; in particular, it computes the similarity coefficient between queries and documents as the probability that a document d_j will be relevant to a query q.

In other words, let r denote the (binary) relevance judgment concerning a set of N documents D with respect to q; the probabilistic model computes the similarity coefficient

$$SC(q, d_j) = P(r = 1|q, d_j), \quad \forall\, d_j \in D$$

The document maximizing such a probability will be ranked as the top result, and other documents will appear in decreasing order of relevance probability.

It must be pointed out that the set of maximally relevant documents is unknown a priori; hence, estimating the above probability is a difficult task. A bootstrapping strategy consists in generating a preliminary probabilistic description of the ideal result set in order to retrieve an initial set of documents. For example, the probability of relevance can be evaluated based on the co-occurrence of terms in the query and documents from the reference document set. At this point, an iterative processing phase can take place (optionally using feedback from the user) with the purpose of improving the probabilistic description of the ideal answer set: based on an intermediate decision as to which of the retrieved documents are confirmed to be relevant, the system refines its description of the ideal answer set, and this process continues until a satisfactory answer set is reached. The following sections provide a detailed discussion of such an iterative algorithm in the context of the binary independence model.

3.4.1 Binary Independence Model

The classic probabilistic retrieval model is the *binary independence model*, where documents (and queries) are represented as binary incidence vectors of terms, i.e., $\mathbf{d}_j = [w_{1j}, \ldots, w_{Mj}]^T$ such that $w_{ij} = 1 \iff d_j$ contains term t_i, and $w_{ij} = 0$ otherwise. The "independence" qualification is due to the fact that the model assumes term occurrences to be independent, a naive assumption that generally works in practice.

Another key assumption in the model is that the distribution of terms t_i in the set $\mathcal{R}$ of documents that are relevant with respect to $\mathbf{q}$ is different from their distribution in the set $\bar{\mathcal{R}}$ of nonrelevant documents. In other words, the probability that a term t_i occurs in a document $\mathbf{d}_j$ (i.e., gets $w_{ij} = 1$) changes based on whether $\mathbf{d}_j$ is relevant or not:

$$P(w_{ij} = 1 | \mathbf{d}_j \in \mathcal{R}, \mathbf{q}) \neq P(w_{ij} = 1 | \mathbf{d}_j \in \bar{\mathcal{R}}, \mathbf{q})$$

In short, we will refer to these two probabilities as $P(t_i | \mathcal{R})$, resp. $P(t_i | \bar{\mathcal{R}})$.

Finally, in the binary independence model, terms that occur in many relevant documents and are absent in many nonrelevant documents should be given greater importance. To account for this, if we refer to the probabilities that $\mathbf{d}_j$ occurs in the relevant, resp. nonrelevant set given query $\mathbf{q}$ as $P(\mathcal{R} | \mathbf{d}_j)$, resp. $P(\bar{\mathcal{R}} | \mathbf{d}_j)$, we might combine the two probabilities so as to define a query/document similarity coefficient. This can be the ratio between the probability that $\mathbf{d}_j$ will be involved in the relevant set $\mathcal{R}$ and the probability that it will be involved in the nonrelevant set $\bar{\mathcal{R}}$:

$$SC(\mathbf{d}_j, \mathbf{q}) = \frac{P(\mathcal{R} | \mathbf{d}_j)}{P(\bar{\mathcal{R}} | \mathbf{d}_j)}$$

Having set the starting point of the probabilistic model, how do we obtain reasonable estimates for the above probabilities in a real-world setting? The following section aims at judiciously rewriting the above equation in order to bootstrap probability estimation.

3.4.2 Bootstrapping Relevance Estimation

What we need now are reasonable estimates for $P(\mathcal{R}|\mathbf{d}_j)$ and $P(\bar{\mathcal{R}}|\mathbf{d}_j)$. To this end, we might consider rewriting $SC(\mathbf{d}_j, \mathbf{q})$ by applying the Bayes rule:

$$SC(\mathbf{d}_j, \mathbf{q}) = \frac{P(\mathcal{R}|\mathbf{d}_j)}{P(\bar{\mathcal{R}}|\mathbf{d}_j)} = \frac{P(\mathbf{d}_j|\mathcal{R})}{P(\mathcal{R})} \times \frac{P(\bar{\mathcal{R}})}{P(\mathbf{d}_j|\bar{\mathcal{R}})}$$

Assuming that $P(\mathcal{R})$ and $P(\bar{\mathcal{R}})$ are constant for each document, we can simplify to:

$$SC(\mathbf{d}_j, \mathbf{q}) = \frac{P(\mathbf{d}_j|\mathcal{R})}{P(\mathbf{d_j}|\bar{\mathcal{R}})}$$

Now, as $\mathbf{d}_j$ is a vector of binary term occurrences with independent distributions, we can rewrite these two terms into something more familiar:

$$P(\mathbf{d}_j|\mathcal{R}) = \prod_{t_i \in D | w_{ij}=1} P(t_i|\mathcal{R}) \quad \text{and} \quad P(\mathbf{d}_j|\bar{\mathcal{R}}) = \prod_{t_i \in D | w_{ij}=1} P(t_i|\bar{\mathcal{R}})$$

Having replaced the latter terms in the SC expression, thus obtaining $\prod_{t_i \in D|w_{ij}=1} \frac{P(t_i|\mathcal{R})}{P(t_i|\bar{\mathcal{R}})}$, we can replace SC by its logarithm for ease of computation. We then get:

$$SC(\mathbf{d}_j, \mathbf{q}) = \log \prod_{t_i \in D|w_{ij}=1} \frac{P(t_i|\mathcal{R})}{P(t_i|\bar{\mathcal{R}})} = \sum_{t_i \in D|w_{ij}=1} \log \frac{P(t_i|\mathcal{R})}{P(t_i|\bar{\mathcal{R}})}$$

Moreover, if we generalize the above expression to all terms in the document set D, we get:

$$SC(\mathbf{d}_j, \mathbf{q}) = \sum_{t_i \in D} \left(\log \frac{P(t_i|\mathcal{R})}{P(t_i|\bar{\mathcal{R}})} - \log \frac{P(\bar{t}_i|\mathcal{R})}{P(\bar{t}_i|\bar{\mathcal{R}})} \right) = \sum_{t_i \in D} \log \frac{P(t_i|\mathcal{R}) \times P(\bar{t}_i|\bar{\mathcal{R}})}{P(t_i|\bar{\mathcal{R}}) \times P(\bar{t}_i|\mathcal{R})}$$

and, considering that probabilities add up to 1, we arrive at the final formulation:

$$SC(\mathbf{d}_j, \mathbf{q}) = \sum_{t_i \in D} \log \frac{P(t_i|\mathcal{R}) \times (1 - P(t_i|\bar{\mathcal{R}}))}{P(t_i|\bar{\mathcal{R}}) \times (1 - P(t_i|\mathcal{R}))}$$

We now have a nice set of probability estimates for the relevance of a given term t_i. However, we still have no means of deciding which documents out of D deserve to be in $\mathcal{R}$ and $\bar{\mathcal{R}}$, respectively. An initial assumption we can make is that $P(t_i|\mathcal{R})$ is a constant, e.g., 0.5; this means that each term has even odds of appearing in a relevant document, as proposed in [98].

Table 3.1 Weight distribution of documents in a set of size N with respect to a term t_i and a query

	Relevant ($\in \mathcal{R}$)	Nonrelevant ($\in \bar{\mathcal{R}}$)	Total ($\in D$)
Documents containing t_i	r_i	$n_i - r_i$	n_i
Documents *not* containing t_i	$R - r_i$	$(N - n_i) - (R - r_i)$	$N - n_i$
Total number of documents	R	$N - R$	N

Also, we can approximate $P(t_i|\bar{\mathcal{R}})$ as the distribution of t_i across the document set. This means that there is no specific reason why terms in the nonrelevant documents should be distributed differently than in the entire document set D. If we call n_i the number of documents initially retrieved from D that contain t_i, we will then be able to say that $P(t_i|\bar{\mathcal{R}}) \sim \log \frac{N-n_i+0.5}{n_i+0.5}$.

The initial set of n_i documents can be obtained either randomly or by using the inverse document frequency (IDF) of each t_i in D as its initial weight; subsequently, the documents can be ranked by SC.

3.4.3 Iterative Refinement and Relevance Feedback

Let us suppose that we can show the initial ranked list to an expert who is willing to provide a *relevance feedback*, i.e., a judgment concerning the relevance of each of the retrieved documents that sets it either in $\mathcal{R}$ or in $\bar{\mathcal{R}}$ [303]. If we consider a term t_i and denote by R the size of $\mathcal{R}$ and by r_i the number of documents in $\mathcal{R}$ that contain t_i, documents will be distributed with respect to t_i according to Table 3.1.

Following the notation of Table 3.1, we can then approximate $P(t_i|\mathcal{R})$ by the frequency of this event in the N documents, i.e., $P(t_i|\mathcal{R}) \sim \frac{r_i}{R}$. Similarly, we can say that $P(t_i|\bar{\mathcal{R}}) \sim \frac{n_i-r_i}{N-R}$; for small values of r_i and R, we can even say that $P(t_i|\bar{\mathcal{R}}) \sim \frac{n_i}{N}$. We remember that $P(t_i|\bar{\mathcal{R}})$ and $P(t_i|\bar{\mathcal{R}})$ add to 1 and, after substitution in the SC expression, we are able to state that

$$SC(\mathbf{d}_j, \mathbf{q}) = \sum_{t_i \in D} \log \frac{r_i(N - R - n_i + r_i)}{(n_i - R_i)(R - r_i)}$$

In other words, $\log \frac{r_i(N-R-n_i+r_i)}{(n_i-R_i)(R-r_i)}$ is the weight w_{ij} given by the probabilistic model to term t_i with respect to document $\mathbf{d}_j$.[1]

At this point, nothing prevents us from repeating this interactive phase with the purpose of improving the probabilistic description of the ideal answer set. By iterating this method as long as relevance feedback progresses (or until a convergence criterion is met), the ranked list of results can be continuously updated.

---

[1] Typically, 0.5 is added to each term to avoid infinite weights.

Table 3.2 A collection of documents about information retrieval

Document	Content
D_1	Information retrieval students work hard
D_2	Hard-working information retrieval students take many classes
D_3	The information retrieval workbook is well written
D_4	The probabilistic model is an information retrieval paradigm
D_5	The Boolean information retrieval model was the first to appear

3.4.4 Evaluation of the Probabilistic Model

The probabilistic model has the advantage of ranking documents according to their decreasing probability of being relevant. In addition, relevance feedback (from the user) concerning the relevance of each result to the query is certainly an advantage. However, this comes at the cost of "guessing" the initial relevance of documents, a task that might not be easy or accurate. Moreover, this model does not take into account the frequency with which a term occurs within one document, and relies on the assumption of the independence of index terms.

The probabilistic model is nowadays a less popular choice than the VSM in state-of-the art IR systems. This is partly due to the impracticality of obtaining relevance feedback and partly to the fact that it was outperformed by the VSM in an evaluation conducted by Salton and Buckley [302]. Indeed, the probabilistic model is penalized by the fact that it only uses IDF, while overlooking aspects such as term frequency (as done in the $tf \times idf$ weighting scheme adopted by the VSM) and document length. A probabilistic model taking these two aspects into account was the Okapi system [296], which achieved an excellent performance at the TREC-8 evaluation campaign.

3.5 Exercises

3.1 Consider the document set in Table 3.2:

- Draw the Boolean term incidence matrix corresponding to the document set.
- Draw the $tf \times idf$ matrix corresponding to the document set.
- Given a simple query, compute the cosine similarity with respect to each document in the document set in the above two cases.

3.2 Recalling Sect. 3.2.1 and given the index characteristics in Table 3.3, propose an order to process the following query by leveraging the frequency of its terms: *(tshirt OR jumper) AND (shorts OR skirt)*.

3.3 Why is the probabilistic model hard to adopt in practice? Why, in contrast, is the vector space model more popular?

Table 3.3 Terms and their
corresponding index
frequency in a small dataset

Term	# Postings
tshirt	187
jumper	232
shorts	31
skirt	448

3.4 The binary independence model assumes that document terms are unrelated and form dimensions in a binary vector space. However, such an assumption is a simplified version of the real-world situation by which terms co-occur at different frequencies and denote semantic relationships undermining such an independence. Why in your opinion does the model hold in practice?

Chapter 4
Classification and Clustering

Abstract Information overload can be addressed through machine learning techniques that organize and categorize large amounts of data. The two main techniques are classification and clustering. Classification is a supervised technique that assigns a class to each data item by performing an initial training phase over a set of human annotated data and then a subsequent phase which applies the classification to the remaining elements. Clustering is an unsupervised technique that does not assume a priori knowledge: data are grouped into categories on the basis of some measure of inherent similarity between instances, in such a way that objects in one cluster are very similar (*compactness* property) and objects in different clusters are different (*separateness* property). Similarity must be calculated in different ways based on the data types it involves, namely categorical, numeric, textual, or mixed data. After the clusters are created, a labeling phase takes care of annotating each cluster with the most descriptive label.

4.1 Addressing Information Overload with Machine Learning

Information overload is one of the main challenges that people, and thus also information retrieval systems, have to face, due to the incredible amount of information needed even for solving simple day-to-day tasks. This chapter focuses on some basic techniques for making sense of such large amounts of information, through an appropriate organization consisting in annotating and structuring the data so as to exploit their similarities [354]. Typical usage scenarios include:

- *document organization*, such as classification of newspaper articles, grouping of scientific papers by topic, and categorization of patents per field of interest;
- *document filtering*, such as recognition of spam or high priority messages in email folders, and selection of documents of interest from incoming streams, such as Really Simple Syndication (RSS) news feeds;
- *Web resources categorization*, such as classifying Web links into hierarchical catalogs.

In all the above cases, given the amount of information involved, the analysis cannot be conducted manually, and thus appropriate automated or semiautomated techniques must be devised.

S. Ceri et al., *Web Information Retrieval*, Data-Centric Systems and Applications,
DOI 10.1007/978-3-642-39314-3_4, © Springer-Verlag Berlin Heidelberg 2013

Data classification helps in accessing information, because it allows users to locate the information of interest through auxiliary navigation structures, based on the computed classes; in this way, they progressively zoom in towards the information of interest through incremental selections, with a limited number of options to be considered at each step. This approach exploits recurrent data patterns that characterize subsets of the data, and enables the users to exploit such patterns as navigation aids on top of the original data.

This chapter presents methods developed within the machine learning discipline, of two main classes:

- *Supervised learning* assumes the availability of a training set of correctly identified observations. *Classification* problems are often modeled as supervised learning problems, with techniques such as naive Bayes, regression, decision trees, and support vector machines.
- *Unsupervised learning*, in contrast, does not assume a priori knowledge but involves grouping data into categories based on some measure of inherent similarity between instances. An example of unsupervised learning is *clustering* (or cluster analysis), which can be performed through partition (e.g., k-means) or hierarchical approaches.

Once the classes (or clusters) are determined, objects can be tagged with the class they belong to. This process is known as *labeling*. In supervised classification, the labeling of the training set is performed by a human actor, while the labeling of the remaining data is performed by the classifier automatically. In contrast, labeling clusters is a complex task that requires the automatic analyzer to select the most representative label(s) for each cluster.

A critical aspect in clustering is the validation of results, which can be based on external criteria (exploiting user's assessment and metrics such as purity, Rand index, and F-measure), or on internal ones (exploiting the dataset itself and metrics such as the silhouette index).

4.2 Classification

Classification is the problem of assigning an object (item or observation) to one or more categories (subpopulations). The individual items or observations are characterized by some quantifiable properties, called *features*; these can be categorical, ordinal, or numerical. An algorithm that implements classification is known as a *classifier*. Some algorithms work only on discrete data, while others also work on continuous values (e.g., real numbers). Classification is based on a training set of data containing observations for which the category is known a priori, as provided by a human analyzer; therefore classification is a supervised learning technique.

In general, given a variable X to be classified and the predefined set of classes Y, training classifiers or labelers involves estimating a function $f : X \rightarrow Y$, which can be formulated as $P(Y|X)$, given a sufficient number of observations x_i of the variable X. Two methods can be applied for addressing the classification problem: generative and discriminative.

- *Generative approaches* assume some function involving $P(X|Y)$ and $P(X)$ and estimate the parameters of both $P(X|Y)$ and $P(X)$ from the training data; then, they use the Bayes theorem to compute $P(Y|X = x_i)$. Notable examples of generative approaches include *naive Bayes* and *regression* classifiers (presented later in this section) as well as *language models* and *hidden Markov models*, discussed in Chap. 5, as they are often applied to natural language processing problems.
- *Discriminative approaches*, in contrast, assume the existence of some function involving $P(Y|X)$ and estimate the parameters of $P(Y|X)$ from the training data; they do not require a model for $P(X)$. In other words, discriminative methods only learn the conditional probability distribution $P(Y|X)$; i.e., they directly estimate posterior probabilities, without attempting to model the underlying probability distributions. Notable examples of discriminative approaches include *decision trees* and *support vector machines* (presented later in this section), as well as *conditional random fields*, discussed in Chap. 5.

Generative methods are based on the joint probability distribution $P(X, Y)$, and are typically more flexible than discriminative models in expressing dependencies in complex learning tasks, while discriminative models can yield superior performance in simple classification tasks that do not require the joint distribution.

4.2.1 Naive Bayes Classifiers

Probabilistic classifiers are based on the Bayes theorem of strong (naive) independence assumption $P(Y|X) = \frac{P(X|Y)P(Y)}{P(X)}$. The posterior probability $P(Y|X)$ is estimated as the prior probability of X generating Y, i.e., $P(X|Y)$, times the likelihood of Y, i.e., $P(Y)$, divided by the evidence for X, i.e., $P(X)$. In the classification context, the probability to be estimated is the probability of an object belonging to a class, given a number n of its features: $P(C|F_1, \ldots, F_n)$.

The naive Bayes classifier can be defined by combining the naive Bayes probability model with a decision rule. A typical choice is to pick the hypothesis that is most probable; therefore, the classification function simply assigns the element with feature values $f_1, \ldots, f_n$ to the most probable class:

$$\text{classify}(f_1, \ldots, f_n) = \text{argmax } P(C = c) \prod_{i=1}^{n} P(F_i = f_i | C = c) \qquad (4.1)$$

Note that the above formulation has a problem: if a class and feature value never occur together in the training set, the corresponding probability estimate will be zero. Since probabilities are multiplied in the formula, a value equal to zero wipes out any other information, too. To solve this problem, the solution is to incorporate in the training a small correction in all probability estimates that removes the zeros.

Fig. 4.1 Logistic function representation

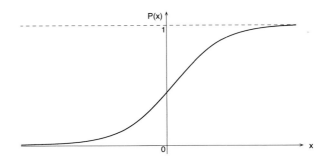

Despite being a very simple technique, naive Bayes is often very effective. If the conditional independence assumption actually holds, a naive Bayes classifier converges more quickly than discriminative models like logistic regression, which in practice means that less training data are needed. Its main disadvantage is its inability to learn interactions between features.

4.2.2 Regression Classifiers

Linear regression classifiers model the relationship between Y and one or more explanatory variables x_i as a linear combination:

$$Y = \beta_1 x_1 + \beta_2 x_2 + \cdots + \beta_{n-1} x_{n-1} + \beta_n x_n + \varepsilon \qquad (4.2)$$

The values of β_i and ε are estimated based on observed data.

In *logistic regression*, the function linking Y and the x_i is not linear but logistic; this is suitable for binomially or multinomially distributed data [167], i.e., categorical data which can assume two (binomial case) or more (multinomial case) possible values. Logistic regression is based on the logistic function (shown in Fig. 4.1), which has the useful property of taking in input any value in $(-\infty, +\infty)$ and producing in output values between 0 and 1.

Logistic regression has a simple probabilistic interpretation and allows models to be updated to reflect new data easily (unlike, e.g., decision trees or support vector machines). Logistic regression approaches are also known as maximum entropy (MaxEnt) techniques, because they are based on the principle that the probability distribution that best represents the current state of knowledge is the one with the largest information-theoretical entropy. Logistic regression can be applied to problems like predicting the presence or absence of a disease given the characteristics of patients, or predicting the outcomes of political elections.

Fig. 4.2 Portion of a decision tree for part-of-speech tagging. A word is decided to be a noun if its predecessor is an adjective preceded by a determiner; otherwise, other properties must be considered to make a decision

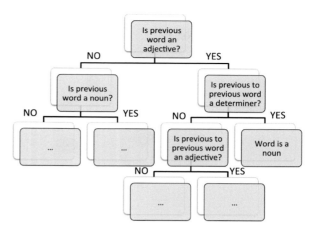

4.2.3 Decision Trees

Decision trees are decision support tools representing possible alternatives as trees with associated probabilities. They can be used for classification by creating a model that predicts the value of a target variable based on several input variables (attributes), in a way similar to logistic regression.

For example, Fig. 4.2 illustrates the portion of a decision tree identifying the part-of-speech tag (noun, verb, adjective, etc.) of a word in a sentence based on the part-of-speech tags of its predecessors.

A classic decision tree algorithm is C4.5 [286], which learns a decision tree from a set of labeled data samples. C4.5 exploits the concept of information entropy as a measure of uncertainty of a random variable: the basic intuition is that at each decision point, the action maximizing the information gain (and therefore minimizing the entropy) associated with the value of the random variable should be taken. In a decision problem, this means associating each decision point with the most discriminative "question" available. For example, in a guessing game where each player selects a person of interest and the other has a limited number of questions to ask in order to find out the person's identity, the best strategy is to ask whether the person of interest is male or female in order to roughly discard half of the candidates. This operation is the one associated with the maximum information gain, as well as the minimum entropy of the resulting population.

Like all recursive algorithms, the C4.5 algorithm defines "base cases," i.e., situations that trivially stop the recursion. The algorithm has three base cases:

- All the samples belong to the same class. In this case, create a leaf corresponding to that class.
- All instances yield zero information gain.
- All instances have the same set of attributes.

C4.5 starts by considering the whole instance set and proceeds recursively as follows:

Fig. 4.3 A graphical illustration of support vector machines. The *decision hyperplane* separates the instances (*black* vs. *white nodes*) of the two classes; its margins are determined by *support vectors*

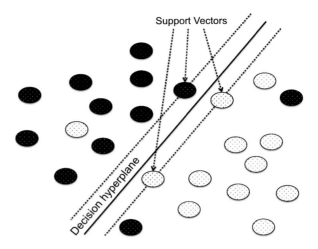

1. Check for base cases that trivially end the recursion;
2. For each attribute, compute the information gain when branching the tree on it;
3. Let a_{best} be the attribute with the highest information gain;
4. Create a decision node that branches based on attribute a_{best};
5. Recurse on the branches produced by attribute a_{best}, add the obtained nodes as children of the current node.

Decision trees are good at handling feature interaction and are insensitive to outliers and to linear separability, i.e., when data points belonging to different classes can be distinguished by means of a linear function. Their main disadvantages are: (i) they do not support online learning, meaning that the tree must be rebuilt if new training samples come in, and (ii) they easily overfit the training data, i.e., they adhere too much to the specificities of training data and do not perform well when applied to previously unseen data. The latter problem can be solved by adopting evolutionary ensemble methods, like random forests or boosted trees [156].

4.2.4 Support Vector Machines

The objective of support vector machines (SVMs) [340] is very simple: given a set of instances belonging to two classes (e.g., *POS* and *NEG*) represented as vectors in a d-dimensional space, find an optimal hyperplane separating the two classes. The term "vector" here refers to the fact that finding such a hyperplane implies finding the instances of each class that minimize the distance from the hyperplane on both "sides" of the hyperplane itself; these are called *support vectors* (see Fig. 4.3).

Formally, the SVM problem consists in finding the decision hyperplane $\mathscr{H}$ described by

$$\mathscr{H} = \mathbf{w} \cdot \mathbf{x} - b = 0 \tag{4.3}$$

where $\mathbf{x}$ represents an instance, $\mathbf{w}$ is a weight vector, and b is a constant. Each instance $\mathbf{x}_i$, $1 \leq i \leq n$ in the dataset must remain within a distance of 1 to the hyperplane, i.e., $\mathbf{w} \cdot \mathbf{x}_i - b \geq 1$ if $x_i \in POS$ and $\mathbf{w} \cdot \mathbf{x}_i - b \leq -1$ otherwise. This can be rewritten as:

$$y_i(\mathbf{w}_i \cdot \mathbf{x}_i - \mathbf{b}) \geq 1, \quad 1 \leq i \leq n \tag{4.4}$$

where y_i represents the label of $\mathbf{x}_i$ in the dataset. The weight vector $\mathbf{w}$ and the constant b are learned by examining all the $\mathbf{x}_i$ instances in the training data and their labels; in particular, the optimal $\mathbf{w}$ can be found by maximizing (4.4) using the method of Lagrange multipliers:

$$\mathbf{w} = \sum_{i=1}^{n} \alpha_i y_i \mathbf{x}_i \tag{4.5}$$

where each instance having a nonzero Lagrange multiplier α_i is a support vector (SV). At this point, b can be computed by taking into account the N_{SV} support vectors:

$$b = \frac{1}{N_{\mathrm{SV}}} \sum_{k=1}^{N_{\mathrm{SV}}} (\mathbf{w} \cdot \mathbf{x}_k - y_k) \tag{4.6}$$

Classifying a new instance $\mathbf{x}_{i+1}$ now simply implies computing $\mathbf{w} \cdot \mathbf{x}_{i+1} - b$.

Clearly, not all classification problems are binary, and not all vector representations allow for a linear separation of the dataset via a single hyperspace. To account for the first issue, the classic solution is to perform one-vs.-all classification, i.e., to train a binary classifier for each class C comparing instances from C against instances of the remaining subclasses (i.e., $\neg C$). The class associated to the highest-scoring binary classifier is selected for the first classification step.

The second issue may be solved by the choice of an accurate feature space, resulting in a linearly separable representation of features; moreover, "soft-margin" SVMs have been designed to tolerate small amounts of mixing for the sake of computation speed or to make the problem separable [355].

SVMs are an extremely efficient technique for supervised classification due to their convergence properties [340].

4.3 Clustering

Clustering is a method for creating groups of objects, called *clusters*, in such a way that objects in one cluster are very similar (*compactness* property) and objects in different clusters are different (*separateness* property). The criteria involved in the clustering process are generally based on a (dis)similarity measure between objects. Data clustering is an unsupervised technique because no a priori labeling of training objects is required and also because it is not necessary to know in advance which

Fig. 4.4 Steps and output of the clustering process

and how many clusters are needed, or what plays a role in forming these clusters [133].

Figure 4.4 summarizes the main steps involved in the clustering process:

1. *Data Processing*. Characterizes data through a set of highly representative and discriminant descriptors (i.e., *features*).
2. *Similarity Function Selection*. Defines how data objects must be compared. The definition of the most effective similarity function for the given context and dataset is a crucial factor for the overall quality of the clustering result.
3. *Cluster Analysis*. Groups data items with a clustering algorithm and according to the chosen similarity function.
4. *Cluster Validation*. Evaluates the produced clusters. The quality of the clusters is evaluated using *validity indexes*.
5. *Cluster Labeling*. Infers highly descriptive and meaningful labels for each generated cluster.

The following subsections overview the techniques that can be applied in each step, with an outlook on advanced issues such as mixed data type management.

4.3.1 Data Processing

Data items are described by typed features, the main data types being numerical, nominal, and textual. *Numerical data* are further described by the degree of quantization in the data (which can be binary, discrete, or continuous) and the value scale. *Nominal data*, also called categorical, can assume a value within a finite enumeration of labels. *Textual data* are sequences of words that compose complex phrases. If a feature used to represent objects is not type-consistent, the referred data is called *mixed data*.

Data clustering algorithms are strictly related to the characteristics of the involved datasets. For that reason, data must be processed in order to grant that the features that will be considered for the clustering are properly normalized. Scale, normalization, and standardization must be applied before performing the actual clustering.

4.3.2 Similarity Function Selection

Clustering groups similar data objects and employs for this purpose a *similarity function* (or *similarity index*) for each data type. Let $\mathbf{x} = (x_1, x_2, \ldots, x_n)$ and

$\mathbf{y} = (y_1, y_2, \ldots, y_n)$ be two n-dimensional data objects. Then the similarity index between $\mathbf{x}$ and $\mathbf{y}$ obviously depends on how close their values are, considering all attributes (features), and is expressed as a function of their attribute values, i.e., $s(\mathbf{x}, \mathbf{y}) = s(x_1, x_2, \ldots, x_n, y_1, y_2, \ldots, y_n)$. The inverse of the similarity function is called the *dissimilarity function*, or *distance function*, defined as $d(\mathbf{x}, \mathbf{y}) = d(x_1, x_2, \ldots, x_n, y_1, y_2, \ldots, y_n)$. Similarity is calculated in different ways based on the data types to which it is applied.

Similarity metrics for *numeric features* have been extensively addressed in the past [112], and the choice of the optimal metric is strictly dependent on the application context. Some of the most used similarity functions are *Euclidean distance*, *Manhattan distance*, and their generalization called *Minkowski distance*, defined as:

$$d = \left(\sum_{i=1}^{n} |x_i - y_i|^p \right)^{1/p} \tag{4.7}$$

where p is the order of the Minkowski distance. Values of $p = 1$ and $p = 2$ yield the Manhattan and the Euclidean distance, respectively.

The most used dissimilarity measure for *categorical data objects* is a well-known measure called *simple matching distance*, also known as *overlap distance*, which produces in output a Boolean value [133]:

$$overlap(x, y) = \begin{cases} 0, & \text{if } x = y \\ 1, & \text{otherwise} \end{cases} \tag{4.8}$$

The dissimilarity between two items is calculated as the sum of the distances between feature value pairs. In realistic settings, the distance function should also take into account the significance of attributes [13].

In many applications, data have descriptors of different types. To address this scenario, two strategies can be adopted:

- *Data Standardization*. Data items are converted into a common type. Typically, nominal values are transformed into integer values. Numeric distance measures are then applied for computing similarity [133]. Another popular method exploits a *binary coding scheme*: a binary value is associated to each of the g categorical classes, and a numerical value is obtained by considering the binary pattern as the representation of an integer value. The inverse approach is the *discretization* of numerical attributes into a finite set of classes. However, discretization leads to loss of information.
- *General Metric Definition*. Feature values are not converted, but a metric for computing similarity over values of different types is defined. The most used distance function of this kind is called the *Heterogeneous Euclidean-Overlap Metric (HEOM) function* [175], which uses a weighted combination of the distance functions specific of the involved types. In particular, it combines an overlap metric for nominal attributes and normalized Euclidean distance for numeric attributes. Several variations of the HEOM function have been defined [352].

4.3.3 Cluster Analysis

Clustering can be seen as an optimization problem where the aim is to allocate in the best way items to clusters. Two main categories of clustering methods exist: *hard* and *soft* (also known as *fuzzy*). In hard clustering each element can belong to only one cluster. In soft clustering, each object can belong to different categories, with a degree of membership that denotes the likelihood of belonging to that cluster.

Partitional clustering is defined as a form of hard clustering that produces partitions of the dataset; in contrast, overlapping clustering describes the case of clusters that admit overlap of objects. *Hierarchical clustering* deals with the case of hierarchies of clusters where objects that belong to a child cluster also belong to the parent cluster [112].

Clusters can be described by characteristic parameters, namely:

- *centroid*, the mean point (middle) of the cluster. The centroid of the mth cluster composed of n objects t_{mi} is defined as:

$$C_m = \frac{\sum_{i=1}^{n} t_{mi}}{n} \qquad (4.9)$$

- *radius*, the square root of the average distance of any point of the cluster from its centroid:

$$R_m = \sqrt{\frac{\sum_{i=1}^{n} (t_{mi} - C_m)^2}{n}} \qquad (4.10)$$

- *diameter*, the square root of the average mean squared distance between all pairs of points in the cluster:

$$D_m = \sqrt{\frac{\sum_{i=1}^{n} \sum_{j=1}^{n} (t_{mi} - t_{mj})^2}{n(n-1)}} \qquad (4.11)$$

Partitional clustering methods divide the dataset into nonoverlapping groups (partitions). The k-means algorithm is one of the most used partitional clustering algorithms [234]. It assumes that the number k of clusters is fixed and known a priori and calculates a representative vector for each cluster, i.e., the *centroid* or mean of the cluster. The centroid may or may not be a member of the original dataset. An intuitive representation of a clustered dataset with the respective centroids is shown in Fig. 4.5.

The algorithm is based on the minimization of an error function, as follows. It starts from a given initial set of k cluster seed points, assumed to be the initial representatives of the clusters. The seed points could be randomly selected or produced through some kind of informed guess. Then, the algorithm proceeds by allocating the remaining objects to the nearest clusters, according to the chosen distance function. At each step, a new set of seed points is determined as the mean (centroid) of each cluster and the membership of the objects to the clusters is updated based

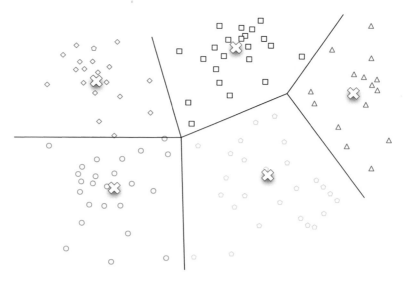

Fig. 4.5 Example of clusters and respective centroids (represented by *X symbols*)

on the error function. This iterative process stops when the error function does not change significantly or the membership of objects to the clusters no longer changes. Notice that due to the presence of local minima in the error function, different starting points may lead to very different final partitions, which could be very far from the globally optimal solution.

Some variants of the *k*-means algorithm exist. The *k-modes* algorithm [174] has been specifically designed for nominal data types, and an integration of *k*-means and *k*-modes, called *k-prototype* [175], has been proposed to cluster mixed (numerical and categorical) data.

Hierarchical clustering methods define hierarchies of nested clusters. Objects that belong to a subcluster also belong to all the clusters containing it. Hierarchical clustering algorithms exploit a *distance matrix* (or dissimilarity matrix), which represents all the pair-wise distances between the objects of the dataset. Hierarchical clustering results are best represented through *dendrograms*, as illustrated in Fig. 4.6. A dendrogram consists of a graph composed of many upside-down U-shaped lines that connect data points in a hierarchical tree of the clusters. The height of each U represents the distance between the two connected data points. The leaf nodes represent objects (individual observations) and the inner nodes represent the clusters to which the data belong, with the height denoting the intergroup dissimilarity. The advantage of dendrograms is that one can cut the diagram with a horizontal line at any height and thus define a clustering level with the corresponding number of clusters (e.g., Fig. 4.6 shows the cuts at 2 and 7 clusters).

Hierarchical algorithms adopt two possible strategies: *divisive* and *agglomerative*.

Fig. 4.6 Example of
dendrogram with cut lines at
2 and 7 clusters

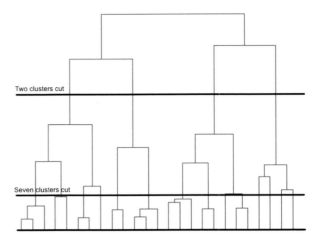

- *Divisive hierarchical clustering* proceeds starting with one large cluster containing all the data points in the dataset and continues splitting it into more and more detailed clusters, with a top-down strategy.
- *Agglomerative hierarchical clustering* starts with elementary clusters, each containing one data point, and continues by merging the clusters, with a bottom-up strategy.

Due to their high computational complexity, which makes them very time-consuming, divisive algorithms are not extensively adopted in applications that deal with large datasets. For that reason we discuss only agglomerative hierarchical algorithms, which are further subdivided into graph methods and geometric methods.

- *Graph methods* represent data as a graph where edges denote a closeness measure between objects (and therefore an ideal cluster would be a clique, i.e., a subgraph such that every two vertices in the subgraph are connected by an edge).
- *Geometric methods* rely on the calculation of distance between nodes.

Various algorithms have been proposed to implement hierarchical clustering efficiently. BIRCH [370] is an efficient algorithm for large numeric datasets, based on a pre-clustering step. Breunig et al. [19] proposed OPTICS, a hierarchical clustering algorithm that can work with data bubbles that contain essential information about compressed data points, where the compression of the dataset is done using BIRCH. Chameleon [191] uses a similarity measure between two clusters based on a dynamic model; two clusters are merged only if the interconnectivity and proximity between them are high, relative to the internal interconnectivity of the clusters and to the closeness of items within the clusters.

Conceptual clustering methods are used for categorical data [128, 218]. These algorithms use conditional probability estimates to define relations between groups or clusters, and they typically employ the concept of link to measure the similarity between data points. The number of links between two points is the number of common neighbors they share. Clusters are merged through hierarchical clustering, which checks the links while merging clusters. Recently, Hsu et al. [172] have proposed a hierarchical clustering algorithm for mixed data based on a new distance representation, called distance hierarchy, which is an extension of the concept hierarchy.

4.3.4 Cluster Validation

Due to the unsupervised nature of the process, the final partition of data produced by clustering algorithms requires some kind of evaluation. The measure of the quality of a clustering scheme is called the *cluster validity index* [133]. Cluster validity can be checked with *external criteria* or *internal criteria*.

External criteria are based on the evaluation of the results with respect to a partial data clustering defined by third parties (humans, systems, or intrinsic data properties), which is chosen as a good example. The most used external criteria are:

- *Purity*. Measured as the normalized count of the number of correctly assigned objects. Bad clusters have purity close to 0, while good ones have purity close to 1. The main limitation of purity is that it cannot be used to evaluate the trade-off between the quality and the number of clusters. Indeed, for a very high number of clusters it is easy to get a high value of purity. In the extreme case of one item per cluster, the purity is always 1.
- *Rand Index*. Measures the accuracy of the clustering, i.e., the percentage of decisions that are correct, by counting false positives and false negatives, considering them with the same weight. It is defined as:

$$RI = \frac{TP + TN}{TP + FP + TN + FN} \qquad (4.12)$$

  where *TP*, *TN*, *FP*, and *FN* are, respectively, the number of true positives, true negatives, false positives, and false negatives.
- *F-Measure*. Also measures the accuracy of the cluster, but it allows one to penalize differently false negatives and false positives, by varying a parameter β. The *F*-measure is an aggregate quality measure of precision and recall, already introduced in Chap. 1:

$$F_\beta = \frac{(\beta^2 + 1)PR}{\beta^2 P + R} \qquad (4.13)$$

  where *P* is the precision and *R* is the recall.

Internal criteria evaluate the results of clustering using the features of the data points. They assume that a good clustering algorithm should produce clusters whose members are close to each other (maximize the *compactness*, i.e., minimize the intra-cluster distance) and well separated (maximize the *separateness*, i.e., maximize the inter-cluster distance). Internal evaluation measures are useful to get some quantitative insight on whether one algorithm performs better than another, but this does not imply that the perceived semantic quality of the output is better from a user perspective.

The *silhouette method* [299] is one of the most used internal criteria, because it provides a measure of both compactness and separateness. Given:

- a set of clusters X_j, with $j = 1, \ldots, k$,
- the i-th member x_{ij} of cluster X_j, where $i = 1, \ldots, m_j$, and m is the number of elements in the cluster,

the silhouette assigns x_{ij} a quality measure called the *silhouette width* defined as:

$$s_{ij} = \frac{b_i - a_i}{\max(a_i, b_i)} \tag{4.14}$$

where

- a_i is the average distance between x_{ij} and all other members in X_j,
- b_i is the minimum average distance of x_{ij} to all the objects of the other clusters.

It can be seen that s_{ij} can assume values between -1 and $+1$, where s_{ij} equal to 1 means that s_{ij} is definitely in the proper cluster.

For hierarchical clustering techniques, the *cophenetic correlation coefficient* is used as a measure of how well the distance in the dendrogram reflects the natural distance (e.g., the Euclidean distance) between the data points. In particular, it expresses the correlation between the dissimilarity matrix P and the cophenetic matrix P_c, where P_c is defined in such a way that the element $P_c(i, j)$ represents the proximity level at which the two data points x_i and x_j are found in the same cluster for the first time.

4.3.5 Labeling

Labeling applies knowledge discovery techniques to define a meaningful and representative label for each cluster. Different techniques are used for textual, categorical, and numerical data. Labeling techniques can be classified as differential or cluster-internal techniques.

Differential cluster labeling compares the relevance of the candidate labels with respect to the cluster under examination and to the other clusters; representative labels are those that are frequent in the current cluster and rare in the others.

For texts, the *mutual information* method is the most used differential cluster labeling technique. Mutual information measures the degree of dependence between two variables. In text clustering, each term is considered as a possible label, and the mutual information associated to the term represents its probability to be chosen as the representative.

In text clustering two variables are defined: C denotes the membership to a cluster ($C = 1$ is a member, $C = 0$ is not a member), and T denotes the presence of a specific term t in the text ($T = 1$ the term is present, $T = 0$ the term is not present). The mutual information of the two variables is defined as:

$$I(C, T) = \sum_{c \in 0,1} \sum_{t \in 0,1} p(C = c, T = t) \log_2 \left(\frac{p(C = c, T = t)}{p(C = c)p(T = t)} \right) \qquad (4.15)$$

where $p(C = 1)$ represents the probability that a randomly selected document is a member of a particular cluster, and $p(C = 0)$ represents the probability that it is not. Similarly, $p(T = 1)$ represents the probability that a randomly selected document contains a given term and $p(T = 0)$ represents the probability that it does not. The joint probability distribution function $p(C, T)$ represents the probability that two events occur simultaneously and is a good indicator of the fact that the term t is a proper representative of the cluster.

Other approaches use a predefined ontology of labels for the labeling process. In [120] the authors combine the standard model-theoretic semantics to partitional clustering. In [132], the authors instead propose an algorithm that integrates the text semantic to the incremental clustering process. The clusters are represented using a semantic histogram that measures the distribution of semantic similarities within each cluster.

Cluster-internal labeling produces labels that only depend on the contents of the current cluster, without considering the characteristics of the other clusters. It is based on simple techniques, such as using the terms that occur frequently in the centroid or in the documents close to it. For numeric data, the label could simply be the average value of the cluster elements; alternatively, numerical attributes can be discretized into labels by using a regular interval subdivision and by assigning a representative value label to each interval. In the case of categorical values, the label could be the most frequent category appearing in the centroid. The disadvantage of these approaches is that they can lead to repeated labels for different clusters.

4.4 Application Scenarios for Clustering

Clustering techniques are applied in several fields. To conclude their overview, we illustrate two important applications: search results clustering and database clustering.

4.4.1 Search Results Clustering

An example of a website that extensively applies clustering techniques to search results is Clusty.com (see Fig. 4.7). Clustering makes it easier to scan results due to the organization in coherent groups, and to discriminate between different meanings of ambiguous terms. It also enables exploratory browsing (e.g., along the paths of the clustering hierarchical trees). Other examples are sites like Google News, where the

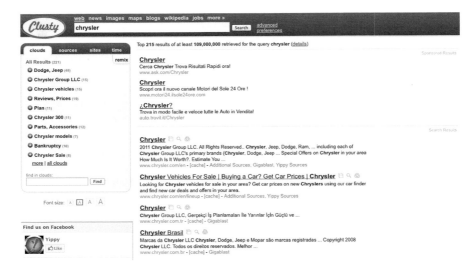

Fig. 4.7 Clusty.com user interface, which clusters Web search results (clusters are visible in the *box on the left-hand side* of the screen)

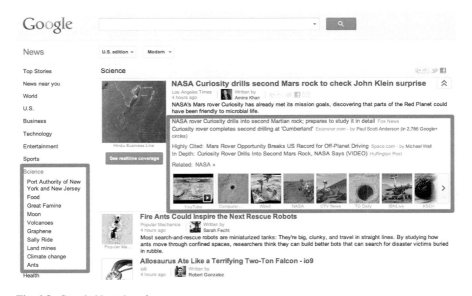

Fig. 4.8 Google News interface

pieces of news are clustered in a hierarchical structure, which is frequently recomputed, as shown in Fig. 4.8. Clusters of items are visible both in categories (left-hand side of the image) and within each main news item in the central part of the image.

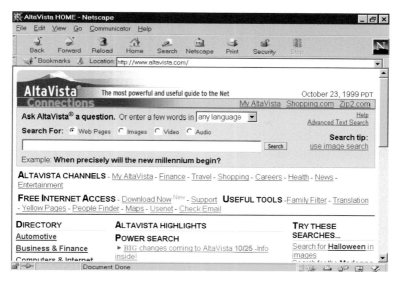

Fig. 4.9 User interface of AltaVista, a popular search engine in the 1990s

Clustering and classification techniques are also applied when computing dynamic facets in faceted search, widely adopted for empowering users in refining their searches. They were also applied by several search engines in the early phases of the Web, such as Yahoo! and AltaVista (a UI dating back to 1999 is shown in Fig. 4.9), for automatically categorizing webpages in directories.

4.4.2 Database Clustering

Relational databases store *tuples*, i.e., strongly typed, structured objects, and clustering can be used in relational databases to identify meaningful subgroups of tuples. In general, it is not possible to directly apply a traditional clustering algorithm to a relational database, because it is first necessary to define which types of objects should be clustered, then which instances should be considered, and finally which attributes to use for clustering. These three steps are a preprocessing stage that the database content must undergo for generating the actual dataset that can be subjected to traditional clustering techniques.

Clustering of multidimensional mixed data occurs when multiple classes of objects and different types of attributes are considered; this problem has been addressed, for instance, in [60], which considered heterogeneous data types and instances from multiple classes. In order to calculate the distance between heterogeneous data types, the work defined a new dissimilarity function that integrates numerical, categorical, and textual attributes. The distance between two elements **x** and **y** is defined as a weighted sum of distances between numeric, categoric, and textual attributes. This allows one to assign different levels of importance to different attributes. The distance is defined as:

$$d(\mathbf{x}, \mathbf{y}) = \sum_{h=1}^{l} \alpha_h (x_h - y_h)^2 + \sum_{i=1}^{m} \beta_i d(x_i, y_i) + \sum_{j=1}^{n} \gamma_j \left(1 - \frac{\mathbf{x}_j^T \mathbf{y}_j}{\|\mathbf{x}_j\| \|\mathbf{y}_j\|} \right) \quad (4.16)$$

where

- the three parts of the function respectively address numerical data, categorical attributes, and textual data. Cosine distance is applied to textual data, where $\mathbf{x}_j$ and $\mathbf{y}_j$ are two documents represented through the vector space model;
- l, m, n are the number of numerical, categorical, and textual attributes, respectively;
- $\alpha_h, \beta_i, \gamma_j$ are the weights associated to a numerical, categorical, and textual attribute, respectively.

For the labeling phase, the problem with mixed types of descriptors is that different types may convey different amounts of information. In particular, textual attributes are the most informative ones, which entails the risk of predominance of textual labels over the others if a single labeling process is applied. To avoid this, labeling over text can be performed independently from numerical and categorical attributes, followed by a final merge step.

While in most cases the labeling process means to find a unique representative label per cluster, in diverse and heterogeneous scenarios like heterogeneous data clustering, it may be useful to represent each cluster with a set of labels, which permits one to produce a better and more intuitive characterization of the clusters.

4.5 Exercises

4.1 Discuss the advantages and disadvantages of generative and discriminative classification approaches.

4.2 Given a set of short text documents, suppose you want to automatically determine: the language of the document, the sentiment of the writer (e.g., positive, negative, or neutral), and a hierarchical categorization of the documents by similarity (with assignment of a descriptive label to each category). Describe and motivate which technique you would use for each of the above issues.

4.3 Use the k-means algorithm and the Euclidean distance to cluster the following 10 bidimensional items into 3 clusters: A = (2, 10), B = (4, 5), C = (3, 2), D = (9, 4), E = (3, 7), F = (5, 5), G = (6, 4), H = (1, 9), I = (1, 2), J = (4, 9).

4.4 Calculate the centroid, diameter, and radius of the three clusters determined in the previous exercise.

Chapter 5
Natural Language Processing for Search

Abstract Unstructured data, i.e., data that has not been created for computer usage, make up about 80 % of the entire amount of digital documents. Most of the time, unstructured data are textual documents written in natural language: clearly, this kind of data is a powerful information source that needs to be handled well. Access to unstructured data may be greatly improved with respect to traditional information retrieval methods by using deep language understanding methods. In this chapter, we provide a brief overview of the relationship between natural language processing and search applications. We describe some machine learning methods that are used for formalizing natural language problems in probabilistic terms. We then discuss the main challenges behind automatic text processing, focusing on question answering as a representative example of the application of various deep text processing techniques.

5.1 Challenges of Natural Language Processing

The Web offers huge amounts of unstructured textual data, i.e., documents that do not have a predefined data model and/or do not fit well into relational tables: home pages, blogs, and product descriptions and reviews are representative of this type of information. In 1998, Merrill Lynch cited estimates that up to 80 % of all potentially usable business information originates in unstructured form [314]. More recently, [148] found that unstructured information accounted for about 70–80 % of all data in organizations and *is growing ten to fifty times as fast as structured data*!

The huge amounts of textual data available on the Internet (currently, the Web offers over 20 billion pages) and company intranets is an opportunity and a challenge to a number of applications. Beyond indexing and search, classic textual operations include text categorization, information extraction, knowledge acquisition, automatic translation and summarization, automatic question answering, text generation, speech interpretation, and human–computer dialog—to cite a few examples.

All such applications, often performed on the results of information retrieval (IR) engines, require a certain level of "understanding" of the natural language in which such (typically unstructured) textual data is written. For this and many other reasons, IR makes extensive use of natural language processing (NLP) methods; natural language interfaces are often regarded as the holy grail of search systems, a fact made

S. Ceri et al., *Web Information Retrieval*, Data-Centric Systems and Applications,
DOI 10.1007/978-3-642-39314-3_5, © Springer-Verlag Berlin Heidelberg 2013

especially true by the recent progress in speech-based search interfaces on mobile systems, which are required to provide meaningful results to spontaneous queries rather than keyword sequences.

5.1.1 Dealing with Ambiguity

A great source of problems arises when devising natural language interfaces to search systems: *ambiguity*. While computer languages are designed by grammars that produce a unique parse for each sentence, natural languages do not benefit from this property (think of the various ways of parsing the sentence "*I saw a man on a hill with a telescope.*").

A closer look shows that the hidden structure of language is ambiguous *at different levels*. Let us consider the sentence "*Fed raises interest rates in effort to control inflation.*" First of all, there is a *morphological* (word-level) ambiguity, as the terms "raises", "interest", and "rates" may be interpreted both as verbs or nouns depending on the context. Moreover, there is a *syntactic* (sentence-level) ambiguity when it comes to attaching "to control inflation" to either "in effort" or "raises interest rates". Finally, there is a *semantic* (meaning-level) ambiguity when interpreting "fed" and "interest", which can both have more than one meaning, only one of which refers to finance.

5.1.2 Leveraging Probability

At the risk of oversimplification, we may identify two complementary ways to process natural language. The first one is *rule-based*: we follow linguistically motivated rules or apply manually acquired resources (e.g., dictionaries) to classify and interpret natural language. In contrast, *statistical* approaches use large volumes of data to drive the inference of patterns and regularities in natural language.

In the processing of natural language, taking probability into account is crucial to rule out unlikely interpretations of ambiguous text. A classic example is "*time flies like an arrow*", where several interpretations are possible, but the one comparing the speed of time to that of an arrow is preferred over, e.g., the interpretation by which insects of the type "time flies" are fond of a particular arrow. Accounting for probability is crucial in applications such as automatic speech recognition, as it allows us to take advantage of the fact that some sequences of words are more likely than others. Similar reasoning occurs when we choose "recognize speech" over "wreck a nice beach", or "let us pray" over "lettuce spray", given a sequence of phones.

For these reasons, machine learning methods are often preferred in the context of NLP when a sufficient amount of representative data exists for the problem, as they join the best of the linguistic and probabilistic worlds. On the one hand, data description often derives from linguistically motivated *feature* extractors (e.g., syntactic parse trees or semantic role labels); on the other, machine learning methods allow for the automated discovery of rules. Next, we focus on machine learning methods which are specifically applied to NLP.

5.2 Modeling Natural Language Tasks with Machine Learning

Many NLP problems can be reduced to the two classic machine learning problems of supervised classification or sequence labeling.

- Supervised classification methods build models from given sets of training examples (with known class label); examples of supervised learning applications include text categorization [311], the classification of question/candidate answer pairs [256], and opinion mining [121]. The most widely adopted classification methods in NLP include naive Bayes, regression, decision trees, and support vector machines: these four methods have been introduced in Chap. 4.
- Given a sequence of observations x_i, sequence labeling methods estimate the most probable sequence of labels y_i "generating" them. Examples of sequence labeling problems include part-of-speech tagging [87], automatic speech recognition [219], and the segmentation of discourse into dialog act markers [284].

We next focus on a specific class of *sequence labeling methods* which play a crucial role in NLP. These methods include language models, hidden Markov models, and conditional random fields.

5.2.1 Language Models

The aim of language modeling is to predict the next word of a text given the previous $n - 1$ words in a probabilistic fashion, i.e., to estimate $P(w_k | w_1 w_2 \cdots w_{k-1})$ (here, $Y = w_k$ and $X = [w_1 w_2 \cdots w_{k-1}]$).

In particular, *n-gram* language models assume a "limited horizon" where

$$P(w_k | w_1 w_2 \cdots w_{k-1}) = P(w_k | w_{k-n} \cdots w_{k-1})$$

i.e., each word only depends on the preceding $n - 1$ words. Specific cases of n-gram models are:

- the *unigram* model, by which $P(w_k | w_1 w_2 \cdots w_{k-1}) = P(w_k)$, and hence words are independent,
- the *bigram* model, by which $P(w_k | w_1 w_2 \cdots w_{k-1}) = P(w_k | w_{k-1})$.

Given such a specification, the problem of finding such probabilities can be formulated as a maximum likelihood estimation (based on a corpus of representative text):

$$P_{MLE}(w_k|w_1 \cdots w_{k-1}) = \frac{C(w_1 \cdots w_k)}{C(w_1 \cdots w_{k-1})}$$

where $C(\cdot)$ represents the number of occurrences in the reference corpus. This method obviously suffers from data sparsity, as for the vast majority of possible n-grams, we get zero probability even in the presence of large corpora. However, "smoothing" solutions exist for this problem; we refer the interested reader to [368].

Unigram language models have been successful in a variety of natural language tasks, such as the estimation of the reading level (or difficulty) of text [94]. Higher order n-gram models have been massively adopted for automatic speech recognition and spoken language understanding [176, 321], and also for text categorization problems [358] such as language identification [80] and authorship attribution [194].

5.2.2 Hidden Markov Models

A *Markov model* is a stochastic model of a *Markov chain*, i.e., a process where the probability distribution of future states of the process depends only on the present state (this is known as the *Markov property*). In other words, a Markov model performs a joint probability estimation of both the current state and the current observation: $P(Y|X) = f(P(X,Y))$. As mentioned in [271], Google's PageRank algorithm is essentially a Markov chain over the graph of the Web.

However, language-related problems often need to account for cases where the current state is not known. A classic example is speech recognition [288], where the current state (i.e., the phones being pronounced, or X realizations) is associated with uncertainty due to possible recognition errors, making it difficult to determine the words (the Y observation) generating the phone. To solve such issues, *hidden Markov models (HMMs)* have been introduced. In an HMM, the state is not directly visible, while state-dependent output is visible. Each state has a probability distribution over the possible output tokens; hence, the sequence of tokens generated by an HMM gives some information about the sequence of states.

Typical applications of HMMs to natural language text occur in tasks where the aim is to label word sequences given their context. For instance, in named entity recognition (see [372]) the labels are named entities such as *location*, *person*, or *organization*. Predicting the label of a word in text is a decision made under the uncertainty about the correctness of the label assigned to previous words.

5.2.3 Conditional Random Fields

Conditional random fields [212] (CRFs) are a form of discriminative modeling that processes evidence from the bottom up. CRFs build the probability $P(Y|X)$ by combining multiple features of the data (X); each attribute a of X fits into a feature function f that associates a with a possible label (typically a positive value if a appears within the data, 0 otherwise). Each feature function f carries a weight representing its strength for the proposed label: a highly positive weight indicates a good association between f and the proposed label, a highly negative weight indicates a negative association with the label, and a value close to 0 indicates that f has little impact on identifying the label.

5.3   Question Answering Systems

In February 2011, an artificial intelligence named Watson participated in the legendary US television quiz game *Jeopardy!*, where a few players compete to find the topic or question underlying a hint. Examples of *Jeopardy!* hints are "This number, one of the first 20, uses only one vowel (4 times!)" (the solution being "seventeen"), or "Sakura cheese from Hokkaido is a soft cheese flavored with leaves from this fruit tree" (the solution being "cherry"). Note that playing such a game is extremely difficult not only for humans, but also (perhaps especially) for computers, as clues are often chosen to be ambiguous or straightforward wordplay, and it is difficult to even detect the topic or expected response type at which they are hinting.

However, the Watson system, designed by IBM, was able to win the game against the two all-time *Jeopardy!* champions. How was this possible? Watson was designed according to state-of-the-art question answering research, blending probabilistic and linguistic approaches to efficiently retrieve answers to the hints. In this section, we will illustrate what question answering is and how state-of-the-art question answering works, and we will discuss its challenges, leading to successful applications such as Watson.

5.3.1   What Is Question Answering?

Automatic question answering (QA) systems return concise answers—sentences or phrases—to questions in natural language. Answers are therefore not relevant documents as in traditional IR, but relevant words, phrases, or sentences. On one hand, QA is interesting from an IR viewpoint as it studies means to satisfy the users' information needs; on the other, the high linguistic complexity of QA systems suggests a need for more advanced natural language techniques, that have been shown to be of limited use for more basic IR tasks, e.g., document retrieval.

QA is a long-lived technology, starting from the need to access small databases using natural language ([316] illustrated 15 such systems as early as 1965) and gradually evolving to its current standards. Current QA systems are characterized by their *open-domain* nature, i.e., by the fact that they are able to accept questions about any topic by accessing a potentially infinite source of evidence. Indeed, most current QA systems use the Web as a source of unlimited, unstructured information to support their answers.

QA research generally divides questions into two different groups: *factoid* questions, i.e., questions where the expected answer can be expressed by a fact or noun (such as a person, a location, or a time expression), and *non-factoid* questions, i.e., questions where the answer format cannot be characterized as precisely. Examples of the latter category are definition questions ("What is rubella?") or why-questions ("Why does the moon turn orange?"), whose answers are not easy to obtain. However, identifying answers of the expected type given a factoid question may not be an easy task when it requires reasoning and information fusion. For instance, answering "Who was president when Barack Obama was born?" requires finding Barack Obama's birth date, then identifying the president during that time; addressing "How many presidents have there been since Barack Obama was born?" requires enumerating the presidents after having found Obama's birth date.

5.3.2 Question Answering Phases

The process of an open-domain QA system is generally composed of three phases (see Fig. 5.1):

1. Question processing, when the question is analyzed and converted into a query for the retrieval of relevant documents;
2. Document retrieval, when the query is submitted to an IR engine, generally using the Web as a document repository;
3. Answer extraction, where relevant documents are analyzed in order to extract and return a ranked list of answers.

Optionally, a fourth phase of answer reranking may be deployed in order to evaluate returned answers and apply additional criteria to improve the final ranking.

1. *Question Processing.* This phase is usually centered around *question classification*, the task that maps a question into one of k expected answer classes. This task is crucial, as it constrains the search space of possible answers and contributes to selecting answer extraction strategies specific to a given answer class. Most accurate question classification systems apply supervised machine learning techniques (see Sect. 5.2), e.g., support vector machines (SVMs) [188] or the Sparse Network of Winnows (SNoW) model [76], where questions are encoded using various lexical, syntactic, and semantic features. It has been shown that the question's syntactic structure contributes remarkably to classification accuracy [225]. In addition, the question processing phase identifies relevant query terms (or keywords) to be submitted to the underlying IR engine.
2. *Document Retrieval.* During document retrieval, the top results returned by the underlying IR engine given the query terms are fetched. The corresponding documents are analyzed in order to *identify sentences*, an operation made difficult

Fig. 5.1 Overview of a
state-of-the-art question
answering system

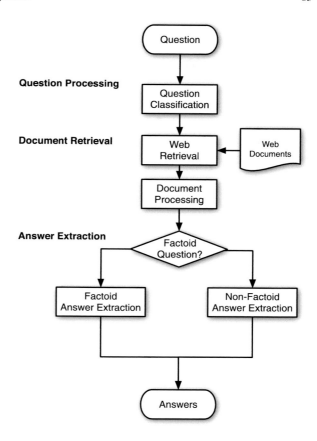

by the heterogeneous structure of Web pages as well as by classic textual segmentation problems (see [40]). Indeed, natural language is rich in punctuation
that does not necessarily constitute sentence delimiters, as in numerical expressions ("0.25"), dates ("19.08.2012", ninth century "B.C."), and denominations
("M.Sc.", "Ph.D.", "M. Obama", "Dr. Ross"). On the other hand, punctuation is
often regarded as unnecessary in Web pages, where the layout is sufficient to understand "sentence" delimiters. Sentence delimitation is therefore often carried
out after "cleaning" the Web document by removing unnecessary content and
possibly identifying textual blocks or passages in the text.

3. *Answer Extraction*. During answer extraction [4], a match is computed between
the query q and each retrieved document sentence s_i in the light of the estimated question class C_q. The *similarity* between q and s_i is usually computed
as a weighted sum of lexical, syntactic, or semantic similarity scores evaluating
how close the question words are to the words in the document sentence [283].
Additionally, specific procedures to the question's class are usually deployed
in order to further constrain the selection of candidate sentences [220]. For in-

stance, named entity recognizers such as LingPipe[1] are used in order to identify instances of persons, locations, or organizations when these are predicted as expected answer types by the question classification modules. Time expressions may be sought using different criteria (e.g., context-free grammars) to answer temporal questions [306].

Obviously, answer extraction criteria for non-factoid questions such as definitions require dedicated modules, because off-the-shelf tools are currently unavailable to identify them. For example, correctly identifying answers to questions about definitions and/or procedures is a challenging task, as these generally include few query terms, making question-answer similarity an insufficient answer extraction criterion. This is the reason why advanced models, such as the ones described in the next section, are often deployed. The final outcome of question/sentence similarity is that the best matching document sentences are identified and ranked based on the above similarity score and returned to the user.

5.3.3 Deep Question Answering

Traditionally, the majority of IR tasks have been solved by means of the bag-of-words approach, optionally augmented by language modeling; the vector space model is an example of this technology. However, some tasks may require the use of more complex semantics and make the above approach less effective and inadequate to perform fine-level textual analysis. For instance, answering some types of questions in automatic QA systems is challenging due to their complex semantics, short texts, and data sparsity. This is the case for definition questions, where queries such as "What are antigens?" and "What is autism?" are characterized by only one keyword, making it difficult to decide on the relevance of answers based on lexical criteria only. The remainder of this section provides a brief overview of a number of "deep" NLP methods that meet the challenge by taking advantage of structural text representations.

In recent years, it has been argued that deeper, more structured linguistic features may be useful in the task of interpreting candidate answers and reranking the results obtained using standard QA techniques. Examples of such features are syntactic parse trees, i.e., trees representing the structure of syntactic relations within a sentence.

Accurate syntactic parse trees can be obtained by using robust, accurate off-the-shelf tools resulting from machine learning approaches; for example, Fig. 5.2 illustrates the syntactic parse tree of a definition as produced by the Stanford probabilistic parser [202]. Here, terminal nodes (nodes having leaves as children) represent each sentence word's part-of-speech tag (DT for determiner, NN for singular noun, etc.), while nodes progressively closer to the root (S) group words in progressively

[1] Available at: alias-i.com/lingpipe.

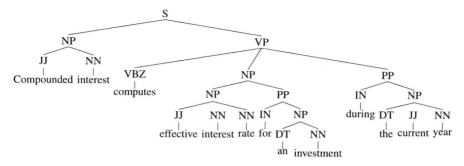

Fig. 5.2 Syntactic parse tree of the sentence "Compounded interest computes effective interest rate for an investment during the current year" according to the Stanford parser [202]

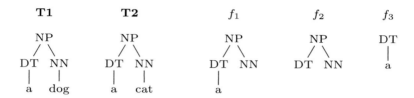

Fig. 5.3 Two syntactic structures with their common subtrees as derived by the tree kernel in [93]

more complex syntactic units. For example, "effective interest rate" is part of a single noun phrase (NP) which in turn is part of a "larger" noun phrase including the prepositional phrase (PP) "for an investment".

An effective way to integrate syntactic structures in machine learning algorithms is the use of tree kernel functions that efficiently enumerate the substructures of tree representations according to different criteria [99]. Two sentences can thus be compared in terms of how many substructures their tree representations have in common. For example, Fig. 5.3 shows two tree structures with their common substructures according to the subset tree kernel defined in [93], which only accepts as valid substructures structures where nodes are either not expanded or fully expanded down to the leaf level.

The overlap between tree substructures is taken as an implicit similarity metric, indeed a much more sophisticated metric than the one estimated by the vector space model. In some sense, we can see tree kernel functions as functions mapping a complex structure such as a tree into a linearly tractable vector that can be processed within a special vector space, accounting not only for the presence/absence of words but also for the presence/absence of specific substructures. Successful applications of different tree kernel functions evaluated over syntactic trees have been reported for many tasks, including question classification [369] and relation extraction [367].

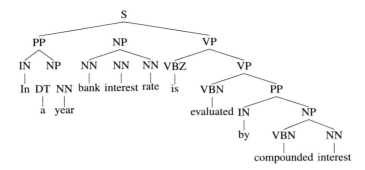

Fig. 5.4 Syntactic parse tree of the sentence "In a year, bank interest rate is evaluated by compounded interest" according to the Stanford parser [202]

5.3.4 Shallow Semantic Structures for Text Representation

There are situations where the information provided by parse trees may prove too sparse: the same concept may produce different, non-matching parses when expressed with different syntax, e.g., in active or passive form. For instance, an alternative definition of compounded interest could read: "In a year, bank interest rate is evaluated by compounded interest", resulting in the parse tree in Fig. 5.4. Despite a number of matching substructures, the overall structure of the latter parse tree does not denote very much overlap with that of Fig. 5.2.

One way to overcome this issue is to try to capture *semantic* relations by adopting representations such as *predicate argument structures*, as proposed in the PropBank project[2] [199]. In a predicate argument structure, words or phrases in a sentence are annotated according to their semantics; for instance, rel denotes a predicate, ARG0 denotes the logical subject of the sentence, ARG1 denotes the logical object, and other, indirect arguments are denoted by ARG2, ARGM-LOC (location), ARGM-TMP (temporal expression), and so on. PropBank annotations can nowadays be accurately obtained using semantic role labeling systems, a review of which can be found in [77]; used as off-the-shelf annotation tools, these reach an accuracy of about 75 % over the average Web text.

Let us consider the semantic role annotation of the previous definitions. The first one would be annotated as "[ARG0 Compounded interest] [rel computes] [ARG1 the effective interest rate for an investment] [ARGM-TMP during the current year]". The second definition would result in: "[ARGM-TMP In a year], [ARG1 the bank interest rate] is [rel evaluated] by [ARG0 compounded interest]".

The key aspects of these structures with respect to syntactic parse trees is twofold. On the one hand, they are much more shallow, in fact resembling shallow parsing (or "chunking"); indeed, they group words together rather than assigning a label to each word that further contributes to a higher level label as in Fig. 5.4. On the other

[2]www.cis.upenn.edu/~ace.

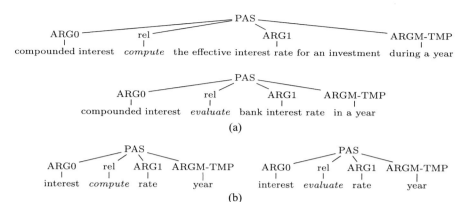

Fig. 5.5 Predicate argument structures of two definitions of the term "compounded interest"

hand, this more shallow notation has the advantage of ignoring a number of fine-grained syntactic distinctions that can intervene in the variations of natural language. For instance, despite the fact that the above sentences are formulated in active and passive form, respectively, their semantic annotation is very similar.

This is more evident if we represent the above annotations via tree structures such as those in Fig. 5.5(a). Figure 5.5(b) reports a yet more compact annotation obtained by automatically extracting noun phrase syntactic heads using, e.g., the algorithm in [92]. By observing these trees, it is easy to note that—despite the different syntaxes—the two definitions they refer to have a very similar *shallow semantic* structure; this in turn suggests the potential efficiency of deploying a sentence similarity function that enumerates and computes common semantic substructures. Indeed, this is what shallow semantic tree kernel functions take care of, as illustrated, e.g., in [257].

5.3.5 Answer Reranking

Rich structural representations such as syntactic and shallow semantic trees seem to be good candidates for combination with classic textual representations such as words, part-of-speech tags, or *n*-grams in order to overcome data sparsity. However, the question arises whether or not this is more effective than the baseline approach of considering "shallow" features alone.

A number of studies prove that the representation of syntax and shallow semantics is beneficial for QA tasks; in [256] for example, supervised discriminative models learn to select (rank) answers using examples of question and answer *pairs*. The pair representation is implicitly provided by combinations of string kernels. Syntactic and shallow semantic tree kernels applied to part-of-speech tag sequences, syntactic parse trees, and predicate argument structures show that it is possible to improve over a bag-of-words model. Such textual features are adopted by a learning model based on SVMs to train binary classifiers deciding whether a ⟨question, answer⟩ pair is correct (i.e., the answer is indeed a valid response to the question) or not.

The findings in [256] and in a number of other studies (e.g., [190, 325]) reveal that the use of structured features is indeed more beneficial for learning classifiers than the use of bag-of-words models. Moreover, it is possible to effectively rerank the answers of a baseline QA system by considering the positive or negative output of the SVM classifier as a criterion to shift incorrect answers down in the original ranking, thus obtaining a reordered version of the original answer set. By using analogous approaches (see, e.g., [142]), IBM's system Watson has been able to select good answer candidates and reach its outstanding performance at the *Jeopardy!* challenge.

5.4 Exercises

5.1 Natural language processing is nowadays a mature technology allowing for spoken and textual interactions with machines using daily language. What existing applications would you qualify as the ones that would draw most benefit from this technology? What applications would benefit from natural language interaction in the near future?

5.2 Is machine learning a "big data" technology? In which ways? Can you name Web resources (search engines, portals, social networks) that adopt machine learning algorithms?

5.3 How can natural language processing techniques such as question answering improve existing information retrieval systems? Can you name search engines that have moved from a traditional, ranked list of relevant documents result set towards a deeper understanding of queries and a more fine-grained analysis of answers? *Hint:* Try searching for people then locations on a state-of-the-art search engine.

Part II
Information Retrieval for the Web

Chapter 6
Search Engines

Abstract In this chapter we discuss the challenges in the design and deployment of search engines, systems that respond to keyword-based queries by extracting results which include pointers to Web pages. The main aspect of search engines is their ability to scale and manage billions of indexed pages dispersed on the Web. We provide an architectural view of the main elements of search engines, and we focus on their main components, namely the *crawling* and *indexing* subsystems.

6.1 The Search Challenge

Estimating the number of Web pages is an interesting search problem. The question can be asked, more or less in these terms, to a semantic search engine, such as WolframAlpha; it would understand the question and report (as of April 2013) that the number of Web pages was 10.82 billion in May 2012, also quoting the source that was used.[1] By accessing that site, one would get a daily estimate of the Web pages on a daily basis, and could read the current estimate (14.65 billion pages as of April 2013). Thus, we can conclude that the Web hosts roughly three pages per Earth inhabitant, and that this number went up from two to three in about one year. Most of the questions that one wants to ask have an answer somewhere in such a huge page space; and search engines are the instruments that help us locate the best candidate pages for these answers.

As the amount of information available on the Web increases, search engines are increasingly faced with the challenging task of answering user queries in a timely and precise manner. Systems such as Google and Bing address these performance and quality challenges by relying upon large, complex infrastructures composed of several, interconnected data centers scattered around the world. The original Google hardware[2] accounted for four machines with a total amount of eight processors, 2 Gb of RAM, and 375 Gb of disk space. In the same year, the overall AltaVista system was known to run on 20 multi-processor machines, all of them having more than 130 Gb of RAM and over 500 Gb of disk space [25].

---

[1] www.worldwidewebsize.com.

[2] http://web.archive.org/web/19990209043945/google.stanford.edu/googlehardware.html.

S. Ceri et al., *Web Information Retrieval*, Data-Centric Systems and Applications,
DOI 10.1007/978-3-642-39314-3_6, © Springer-Verlag Berlin Heidelberg 2013

These early systems are hardly comparable to today's installations. Google stopped reporting the number of indexed documents in August 2005, when Yahoo! claimed[3] an index size of 19 billion resources. In a 2007 paper, Baeza-Yates et al. [29] estimated that the number of servers required by a search engine to handle thousands of queries per second reach the order of millions. No official data about the current situation is available, as the big players keep hardware-related information hidden, so the actual figures are the subject for rumors or rough estimations. It is said that, as of 2012, Google data centers deployed around the world accounted for around two million machines. An interesting insight comes from a recent study that tried to estimate the environmental impact of Google's infrastructure, which is estimated to be around 0.0003 kWh (or 1 kJ) of energy per search query.[4] Google claimed in August 2012 that the company's engine answers 100 billion searches per month, roughly 20 searches per Earth inhabitant. These impressive numbers are the result of a history of developments which spans the last two decades.

6.2 A Brief History of Search Engines

The first search system, Archie, was developed in 1990 by Alan Emtage, a student at McGill University in Montreal, and used a keyword-based filter on a database of filenames. At that time, the World Wide Web was not invented yet, as the first Web page went online in August 1991 at CERN, from an idea of Tim Berners-Lee. The first robot, a system that autonomously searches the web for information, was created in June 1993, when Matthew Gray introduced the World Wide Web Wanderer, a system whose purpose was to count active Web servers. The ingredients of search engine technology were thus at play about 20 years ago, and the technology then developed very rapidly.

The *first generation of search engines* includes systems such as Excite, Lycos, and AltaVista, developed between 1994 and 1997. These systems were ranking search results using classic models of information retrieval, by considering Web pages as textual files and applying the methods discussed in Chap. 3 (specifically, the vector space model). The main limitation of this method was the lack of relevance given to a page because of its popularity; this factor was first taken into account by Lycos, which in 1996 used the *link voting principle*, giving higher popularity to the sites which were most linked by other sites.

This idea evolved into the development of the *second generation of search engines*, marked by the advent of Google, which used a more sophisticated measure of page popularity, in which links from pages with high reputation give higher relevance to pages. This idea is at the basis of link analysis methods, such as the PageRank and HITS algorithms, reviewed in Sects. 7.3 and 7.4, respectively. The mathematical solutions provided by such methods are typically complemented by other

---

[3]http://www.ysearchblog.com/archives/000172.html, accessible through the Web archive at http://web.archive.org/.

[4]http://googleblog.blogspot.it/2009/01/powering-google-search.html.

factors, including the analysis of *click-through data* (which search results are actually selected by people) or of *anchor text* (the text that refers to linked pages).

While there is a clear demarcation line between the first and second generation of search systems, post-link analysis technologies gave rise to what is often referred to as the *third generation of search engines*. However, it is not so clear how to describe these systems, as they exhibit a mix of features, including the ability to perform *semantic search*, but also the ability to integrate with *social networks*, a greater dependency on the *local context* of the search query, and the ability to support *multimedia data*, such as videos. These aspects are described in the third part of the book. The major players of the second generation are progressively including all such features within their products, typically through company and technology acquisitions.

Google is by large the most popular search engine, so it is interesting to describe its history. Its origin can be traced to Stanford's Ph.D. students Larry Page and Sergey Brin, who developed PageRank and started Google in 1998. Their success story started rather traditionally when they received an initially limited seed funding and then a generous round of venture capital; then in 1999, Google became a partner of AOL and Yahoo!. Google soon developed the powerful advertising systems AdWords (in 2002) and AdSense (in 2003), and went public on the Nasdaq market in August 2004. The strength of Google derives also from the complementary services which are offered together with the search engine, including support for email, maps, document sharing, calendars, and so on; in 2007, Google bought YouTube and started inserting videos into search results. In 2010, Google bought Freebase and since then has moved in the direction of adding semantics to search results, which resulted in the introduction, in May 2012, of the Knowledge Graph, a way to semantically describe the properties of searched objects (see Chap. 12).

The main historical competitors of Google are Yahoo! and Bing. *Yahoo!* was founded in 1994 by David Filo and Jerry Yang, also from Stanford, and initially was focused on providing access to websites organized as a well-classified directory of resources. The company used Google for Web search, and developed a proprietary search technology in 2004, which then grew up to become the second in share after Google. Since 2009, the Yahoo! website adopted Microsoft's Bing search engine, and Yahoo! is no longer competing on search. The main strength of Yahoo! is offering platforms for social content sharing, such as Flickr (a photo sharing site) and Delicious (a social benchmarking site).

Microsoft MSN Search was available since 1998, but it became a competitor of Google in the search engine market around 2004; then the company announced the products Live Search (in 2006) and *Bing* (in 2009). Bing introduced a number of innovations, including the ability to generate search suggestions while queries are entered or directly in the result set; it makes use of semantic technology stemming from the acquisition of the Powerset company (in 2008). Bing is strongly integrated with social networks and extracts data from Facebook and Twitter. As of the Spring of 2013, Google and Bing are the major search engine competitors, while a good share of search engine traffic (over 50 % of the queries written in Chinese) is controlled by Baidu, a search engine for the Chinese language.

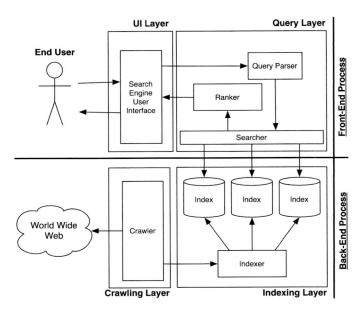

Fig. 6.1 The architecture of a Web search engine

This short and by no means exhaustive snapshot on the recent history of search engines does no justice to the huge number of smaller companies and systems which marked the history of search engines. It shows that mastering Web search is the current most compelling challenge of information technology.

6.3 Architecture and Components

Much like classic search engines (see Chap. 1), Web search engines are roughly composed of a front-end process and a back-end process [21], as shown in Fig. 6.1.

The front-end process involves the *user interface* and *searcher* layers. Upon the submission of a user query, it is devoted to (i) parsing and understanding of the query, (ii) retrieval of relevant results from the search engine indexes, (iii) ranking of the matching results, and (iv) interaction with the user. Given the breadth of such topics, we address them in dedicated chapters of the books (Chap. 2 for parsing of textual queries, Chap. 7 for link analysis and ranking algorithms, and Chap. 14 for search user interfaces).

The back-end process, which is the focus of this chapter, involves the *indexing* and the *crawling* layers, and it is in charge of guaranteeing that the Web search engines are able to handle hundreds of thousands of queries per second, while offering sub-second query processing time and an updated view on the current status of the available Web resources. To cope with such tight constraints, Web search engines constantly crawl the Web, and build indexes that contain a large number of accurately represented resources. Crawling and indexing are described in the next two sections.

6.4 Crawling

In order for a search engine to provide accurate responses to users, Web resources must be identified, analyzed, and cataloged in a timely and exhaustive manner. This content exploration activity, called *Web crawling*, is an automated process that analyzes resources in order to extract information and the link structure that interconnects them. The computer programs in charge of traversing the Web to discover and retrieve new or updated pages are typically referred to as *crawlers*; they can also be called *spiders*, *robots*, *worms*, *walkers*, etc. Crawlers have many purposes (e.g., Web analysis, email address gathering, etc.), although their main one is related to the discovery and indexing of pages for Web search engines. The goal of this section is to describe the main design principles of Web crawlers, and the issues that must be faced for their successful implementation.

The whole set of resources accessible from the Web can be roughly classified into three categories:

- *Web pages*, *textual* documents written in the *HyperText Markup Language* (HTML), using a charset encoding system (US-ASCII, UTF-8, etc.), and containing information and *hypertextual links*;
- *Multimedia* resources, which are *binary* objects such as images, videos, PDFs, Microsoft Word documents, etc., which require a dedicated player or viewer in order to access their contents;
- *Scripts*, the source code of programs that can be executed by a Web client in order to provide dynamic behavior to a Web page.

Resources can be distinguished in *physical* form, i.e., statically defined files hosted by a Web server, or *logical* form, i.e., dynamically produced by a server-side program, and are described by a *uniform resource locator* (URL), a string that contains the information required to access their actual content. For instance, the URL below says that the file siteindex.html, contained in the Consortium directory, can be retrieved from the WWW server running on the domain name w3.org (or, alternatively, at the IP address 128.30.52.37) on the port number 80.

```
http://www.w3.org:80/Consortium/siteindex.html
```

To retrieve the content of a resource, crawlers address a network request to the hosting server using the *HyperText Transfer Protocol* (HTTP). The request must be encoded in a predetermined format, exemplified below; besides the URL of the requested resource, it contains information about the protocol's method to use (GET) and the requesting *user agent*, i.e., the software that interacts with the server on behalf of a user. The user agent information is used by the server to identify the class of connecting clients and their characteristics, in order to enact personalized content provisioning and distribution policies.

```
GET /Consortium/siteindex.html HTTP/1.1
Host: www.w3.org
User-Agent: Mozilla/5.0 Chrome/23.0.1271.101
```

In order to access the resource's content, the crawler has first to *resolve* the server hostname into an IP address, so as to be able to contact it using the IP protocol. The mapping from name to address is done using a distributed database of mappings maintained by known servers offering a *Domain Name Service* (DNS).

Upon the arrival of an HTTP request, the Web server locates the requested *physical* resource (or invokes an application to produce a *logical* one) and returns its content to the requesting client, together with some HTTP response header metadata which might be useful to assist the enactment of proper content analysis and indexing processes (e.g., the timestamp of the last update of the resource).

```
HTTP/1.1 200 OK
Content-Type: text/html; charset=utf-8
```

In the above HTTP response, for instance, the server notifies the client about the MIME type of the resource, the adopted character encoding system, and the overall length of the page in bytes.

6.4.1 Crawling Process

Crawlers discover new resources by recursively traversing the Web. Figure 6.2 sketches the typical steps of a crawling process. The crawler picks the resources to fetch from the URL *frontier*, i.e., a processing queue that contains the URLs corresponding to the resources that have yet to be processed by the crawler. When the crawling process starts up, the URL *frontier* contains an initial set of *seed* URLs; the crawler picks one URL and downloads the associated resource, which is then analyzed.

The analysis aims at extracting all the information which might be relevant for crawling and indexing purposes. For instance, when the fetched resource is a Web page, the analysis task parses the HTML page so as to extract the contained *text* to be later processed and provided to the textual indexer for textual analysis and indexing. The analysis also retrieves *links* pointing to other resources, which are then added to the URL *frontier*. Indeed, different resource types call for alternative analysis methods: for instance, if the downloaded resource is a video, then a video analysis algorithm might be used to extract indexable features (see Chap. 13).

When a resource is fetched and analyzed, it can be either removed from the *frontier*, or added back for future fetching. Although conceptually simple, the recursive traversal of the Web graph is challenged by several practical requirements [241] of a Web crawling system, listed as follows:

Fig. 6.2 The Web crawling
process

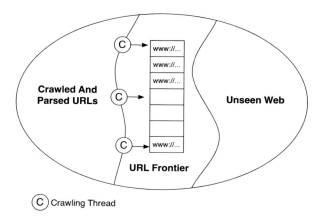

- *Performance*: A crawler should be able to make efficient use of the available system resources (processors, storage, and network bandwidth), possibly by distributing the crawling tasks among several nodes.
- *Scalability*: A crawler should be designed to allow the scaling up of the crawling rate (e.g., number of pages fetched per second) by adding extra machines and bandwidth. Note that *content* scalability (the expected size of the resource collection to fetch) is an implicitly addressed requirement, as the scale of the Web (hundred of billions of resources, petabytes of information created every day) calls for specific implementation solutions (e.g., the usage of 64- or 128-bit address space for memories and disk sectors).
- *Quality*: As not all the resources on the Web are useful for indexing, exploration and, ultimately, query purposes, the crawler should be able to identify "useful" pages, favoring their prompt retrieval and their continuous update. Quality is a subtle requirement, as it entails properties of the crawled data which are application dependent: for instance, quality might be driven by the content of a resource, by its geographical location, its popularity, similarity to the query, etc. Note that the quality of a crawler is also defined by its ability to spot *duplicate* contents, i.e., resources which are addressed by multiple URLs but that contain the same information.
- *Freshness*: Web contents are continuously changing; therefore, crawlers should guarantee that search engine indexes contain the most recent version of a given resource, possibly by visiting resources with a frequency that approximates their change rate.
- *Extensibility*: To cope with the evolving nature of the Web, crawlers should be designed in a modular way, so as to dynamically include new content analysis components, graph traversal functionalities, fetching protocols, etc.
- *Robustness* to *Spider Traps*: A crawler must be able to deal with malicious Web servers that try to trick the crawler into fetching an infinite number of pages in a particular domain. Traps can be involuntary: for instance, a symbolic link within a file system can create a cycle. Other traps are introduced intentionally: for instance, a dynamic website could generate a (potentially) infinite number of items

(e.g., calendar widgets) which, in turn, could create a (potentially) infinite number of pages; website administrators may create traps to discriminate human Web users from automatic processes, to identify new crawlers, or to block "bad" ones (e.g., spammers and crawlers).

- *Politeness*: Web crawlers are automated processes that could severely hamper the performance of a Web server, by performing resource requests at higher rates than humans are capable of. Moreover, crawlers could access the whole set of resources available on a given server, regardless of access limitations expressed by the content owners. Instead, crawlers must comply with the policies defined by the owner of a Web server, both in terms of access policies (see Sect. 6.4.7) and crawling rate. The explicit and consistent identification of the crawler through HTTP *user agents* headers is also considered a fair practice.

Many of the above requirements conflict, typically when they involve the trade-off between *performance/scalability* and *quality/freshness*. The "art and craft" of a successful Web search engine is about identifying the best trade-offs in satisfying all the requirements, and this is the playing field of all the main big players in the search industry.

6.4.2 Architecture of Web Crawlers

In their seminal work, Page and Brin [64] describe the early implementation of the Google Web crawler as a distributed system consisting of five functional components running in different processes. A single *URL server*, fed by a URL database, supplied a set of crawlers, each running as a single thread on a dedicated processing node, and using asynchronous I/O to fetch data from up to 300 Web servers in parallel. Once fetched, documents were sent to a *store server* responsible for the identification, compression, and storage of resources into a repository; the *indexer* process was responsible for parsing the Web page in order to extract links, which were then turned into absolute URLs and stored in the *URL database*. As described in [64], typically three to four crawler machines were used, so the entire system required between four and eight machines.

The crawler *Mercator* [165], adopted by the AltaVista search engine, inspired the implementation of a number of research and commercial crawlers; we will use it as reference. Figure 6.3 describes its organization.

While the Google crawler uses single-threaded crawling processes and asynchronous I/O to perform multiple downloads in parallel, the *Mercator* crawler uses a multi-threaded process in which each thread performs synchronous I/O. Crawling is performed by hundreds of threads, called *workers*, that may be run in a single process, or be partitioned among multiple processes running at different nodes of a distributed system.

Each worker repeatedly performs the steps of the crawling process described in Fig. 6.2. The first step in the process is to pull a new absolute URL from the *frontier*: a *DNS resolution* module determines the Web server from which to fetch the

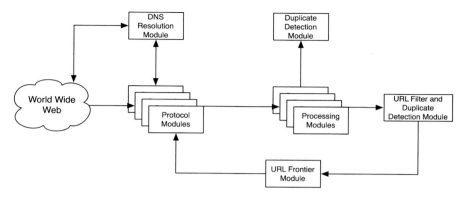

Fig. 6.3 The architecture of the *Mercator* Web search crawler

page specified by the URL and, according to the network protocol (HTTP, FTP, etc.) defined in the URL, the worker invokes the most appropriate *protocol module*. The protocol modules to be used in a crawl are specified in a user-supplied configuration file, and can be dynamically loaded at the start of the crawl. As suggested by the tiling of the protocol modules in Fig. 6.3, there is a separate instance of each protocol module per thread, which allows each thread to access local data without any synchronization. Upon invocation, the *protocol module* downloads the resource, and it signals that the resource is ready for analysis; the *worker thread* applies a *Content Duplicate Detection* test to determine whether this document (associated with a different URL) has been seen before. If so, the document is not processed any further.

Based on the downloaded resource's MIME type, the worker invokes the *processing module* devoted to its analysis. A processing module is an abstraction for processing downloaded resources, for instance, extracting links from HTML pages, and counting the tags found in HTML pages. Like protocol modules, there is a separate instance of each processing module per thread. In general, processing modules may have side effects on the state of the crawler, as well as on their own internal state. The processing module for the MIME type text/html extracts the text and the links contained in the HTML page. The text (possibly stripped from all the HTML tags) is passed on to the indexer;[5] each extracted link is converted into an absolute URL and goes through a series of tests to determine whether the link should be added to the URL frontier.

As the crawling of the Web could take weeks (or months), it is important to periodically check the status of crawlers and consolidate the state of the crawling process into a *checkpoint* that can be used for restoring in case of failures. This "housekeeping" task is performed by a dedicated thread that periodically verifies the status of the crawler by analyzing the log produced by the crawler (URLs crawled, frontier

---

[5]The links information extracted from an HTML page can also be passed on to the indexer for use in ranking functions based on link analysis (see Chap. 7).

size, etc.). According to this analysis, the control thread may decide to terminate the crawl or to create a checkpoint.

6.4.3 DNS Resolution and URL Filtering

Before downloading a resource, the Web crawler must know the actual Web server on the Internet that hosts that resource. As nodes on the Internet are identified by IP addresses, the *Domain Name Service (DNS) resolution* component is in charge of extracting the hostname (e.g., en.wikipedia.com) from an absolute URL (e.g., http://en.wikipedia.org/wiki/Main_Page) and mapping it into an IP address (e.g., 212.52.82.27). This mapping operation is known to be a major bottleneck for Web crawlers, as it requires the interaction with a distributed DNS server infrastructure [204] that might entail multiple requests across the Internet and, ultimately, latencies in the order of seconds. In addition, DNS lookup operations are typically managed by the underlying systems as synchronized methods, thus eliminating the performance boost provided by the use of several crawling threads. To circumvent those problems, Web crawlers typically use custom-made DNS client libraries which, besides providing support for multi-threading, implement caching systems that store the results of DNS lookups for recently visited URLs.

The subsequent *URL filtering* test determines whether the extracted URL should be excluded from the frontier based on user-supplied rules (e.g., to exclude certain domains, say, all .com URLs) or Web server-specific access policies (see Sect. 6.4.7). If the URL passes the filter, the worker performs a *URL normalization* step, in which relative URLs are transformed into absolute URLs. Finally, the worker performs a *URL duplicate elimination* test which checks if the URL has been seen before, namely, if it is in the URL frontier or has already been downloaded; if the test succeeds, then the URL is added to the frontier, and it is assigned a *priority* in the fetching queue. Section 6.4.6 will provide more details about this priority queue.

6.4.4 Duplicate Elimination

Many resources on the Web are available under multiple URLs or mirrored on multiple servers, thus causing any Web crawler to download (and analyze) the same resource's contents multiple times. The purpose of the *content-seen* test is to decide if the resource has already been processed by the system. By some estimates, as many as 40 % of the pages on the Web are duplicates of other pages [241].

The content-seen test can potentially be a very expensive operation, as a similarity comparison should be performed for every possible pair of candidates, requiring a (nearly) quadratic number of pairwise comparisons.[6] Moreover, the similarity test clearly cannot be performed on raw, downloaded resources.

A simple solution to the resources storage problem is the creation of a *document fingerprint* database, storing a succinct (e.g., 64-bit checksum) digest of the contents of each downloaded resource, to use for exact duplicate detection. Then, approaches are needed to improve runtime performance by reducing the number of pairwise comparisons, for instance by partitioning (or clustering) resources into smaller subsets. Under the assumption that duplicate documents appear only within the same partition, it is then sufficient to compare all resource pairs within each partition.

Unfortunately, fingerprint-based approaches fail to capture the presence of *near* (or *fuzzy*) duplicates, i.e., resources that differ only by a slight portion of their contents. *Shingling* is an efficient technique, introduced by Broder et al. [66], for Web page similarity calculation: it relies on the idea that pages can be represented by *sketches*, i.e., canonical sequences of tokens calculated by ignoring HTML tags and applying typical text analysis operations. A contiguous subsequence of size w of tokens in the document is called a *shingle*. Two documents that are near duplicates would feature similar sets of shingles, and the *resemblance* of two pages can then be calculated using the Jaccard coefficient.

6.4.5 Distribution and Parallelization

Computation distribution is an essential and very effective scaling strategy, which can also be adopted in the context of crawling systems. The *Mercator* reference architecture natively supports crawling distribution, which can be easily achieved by employing several nodes, each running dedicated worker threads. Distribution can be performed according to several *splitting* strategies. For instance, crawler nodes can be assigned to specific sets of hosts, possibly assigned according to geographical proximity [28], minimizing the communication latency; alternatively, nodes can be partitioned according to the type of resources to be fetched, in order to properly size the hardware and software required to analyze them.

The main drawback of a distributed architecture stems from the need for co-ordinating several Web servers. A parallel crawling system requires a policy for assigning the URLs that are discovered by each node, as the node that discovers a new URL may not be the one in charge of downloading it [29]. The goal is to avoid multiple downloads and balance the overall computational load. This is achieved by including in the reference architecture a *host splitter* module, which is in charge of dispatching URLs that survive the *URL filter* test to a given crawler node. On the other hand, the local URL repository used by the *URL duplicate elimination* test

---

[6]The total runtime complexity of iterative duplicate detection is $O(n^2)$ for a given candidate set C^T that contains candidates of type T [265].

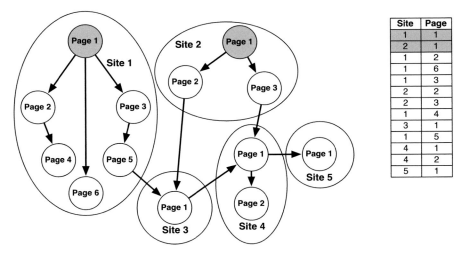

Site	Page
1	1
2	1
1	2
1	6
1	3
2	2
2	3
1	4
3	1
1	5
4	1
4	2
5	1

Fig. 6.4 Example of breadth-first traversal of the Web graph to populate the URL frontier. The initial set of pages is highlighted in *gray*

should be complemented with information shared by the other distributed crawling nodes. Moreover, the *content duplicate detection* module must be enhanced to guarantee that the same (or highly similar) content appearing on different Web servers is not analyzed by different crawlers; this can be achieved in several ways, from remote calls to each processing node to test the uniqueness of the resource (a solution that can quickly saturate the communication links of the system), to distributed hash tables, like the ones adopted in peer-to-peer systems.

6.4.6 Maintenance of the URL Frontier

The *URL frontier* is the data structure that contains all the URLs that will be next downloaded by the crawler. The order by which URLs are served by the frontier is very important, and it is typically bound to the following constraints:

- *Politeness*: Only one connection can be open at a time to any host, and a waiting time of a few seconds must occur between successive requests to a host;
- *Relevance*: High-quality pages (e.g., Wikipedia articles) or pages that frequently change (e.g., news websites) should be prioritized for frequent crawling.

Breadth-first or depth-first traversal strategies, shown in Figs. 6.4 and 6.5, are simple to implement but violate the previous constraints. Using a standard FIFO queue (in which elements are dequeued in the order they were enqueued), the breadth-first strategy can be implemented by adding all pages linked by the current page in the URL frontier: the coverage will be wide but shallow, and a Web server can be bombarded with many rapid requests. On the other hand, a depth-first

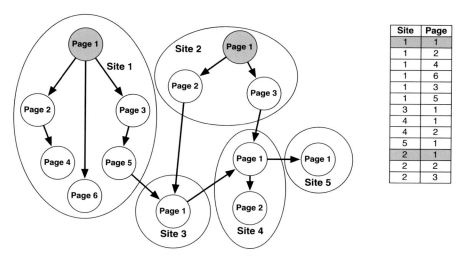

Site	Page
1	1
1	2
1	4
1	6
1	3
1	5
3	1
4	1
4	2
5	1
2	1
2	2
2	3

Fig. 6.5 Example of depth-first traversal of the Web graph to populate the URL frontier. The initial set of pages is highlighted in *gray*

strategy will follow the first link on the current page, doing the same on the resulting pages: the traversal will be very deep, but narrow, and, due to possible reference locality (e.g., URLs that link to other URLs at the same host), there might also be a politeness problem. To comply with these restrictions, it is customary to maintain a queue of websites, and for each website a queue of pages to download.

Note that the naive versions of both strategies do not consider relevance and, therefore, do not prioritize fetches according to the importance of the resource to crawl. However, assuming the seed URLs to be associated with important or popular resources, a crawler that downloads contents in breadth-first search order discovers the highest quality resources during the early stages of the crawl. As the crawl progresses, the quality of the downloaded resources deteriorates [262].

A popular way of estimating the importance of a page is by calculating its relevance, typically according to some link analysis measure. Several crawling strategies have been proposed over the years [29], each with its own advantages and disadvantages.

The *backlink counts* order resources according to the highest number of links pointing to them, so the next resource to be crawled is the most linked from the pages already downloaded [90]. *PageRank*[7] uses the score of a page as a metric of its importance: batch analysis can estimate the PageRank of all the resources in the frontier, so that the next ones to download are the ones with the highest estimated score [90]. The *larger-sites-first* strategy [78] tries to avoid having too many pending

---

[7]PageRank is a page relevance measure that will be described in Chap. 7.

pages in any website,[8] by using the number of uncrawled pages found so far as the priority for picking a website. Strategies can also exploit information extracted from previous crawling activities; history-aware strategies can deal with the pages found in the current crawl which were not found in the previous ones, typically by assigning them an estimated relevance score.

Reference [29] provides an empirical evaluation of the performance provided by several crawling strategies, showing that history-aware strategies are typically very good, and that the breadth-first strategies have a bad performance. While very simple, the *larger-sites-first* strategy provides a performance similar to that of the history-aware ones; strategies based on *PageRank* suffer from severe computational penalties due to repeated computations of the PageRank scores, and their performance is no better than that of simpler strategies.

6.4.7 Crawling Directives

The URL filtering mechanism provides a customizable way to control the set of URLs that are downloaded. Besides crawler-specific rules, website owners can instruct crawlers about the resources that should or should not be included in the search engine index.

The explicit exclusion of resources can be motivated, for instance, by the belief that (part of) the contents hosted by the website can be misleading or irrelevant to the categorization of the site as a whole. On the other hand, site owners might want to enforce the copyright on their contents, with the intent of preserving their value. For instance, several websites employ a business model based on advertisement: having search engines to provide meaningful previews on their search result pages would inevitably lead to less traffic and, ultimately, to less revenues.

The *Robots Exclusion Protocol* [207] is the standard mean for site owners to instruct Web robots about which portions of their websites should be excluded from crawling. This is done by placing a file with the name *robots.txt* at the root of the URL hierarchy at the site. The *robots.txt* is a text file that contains one or more records. Each record defines instructions directed to one or more `User-agent` about one or more URL prefixes that need to be excluded. For instance, the record below instructs all crawlers to avoid visiting the contents of the `/cgi-bin/` and the `/tmp/` folders.

```
User-agent: *
Disallow: /cgi-bin/
Disallow: /tmp/
```

However, the record below specifies that the only crawler allowed to access the website is `Google`, and that the whole set of contents is available.

---

[8]Large websites might be penalized because of the *politeness* constraint, which might force their resources to be fetched last.

```
User-agent: Google
Disallow:

User-agent: *
Disallow: /
```

Extended versions of the Robots Exclusion Protocol include additional commands. For instance, the `request rate` parameter is used to instruct the crawler about the number of seconds to delay between requests, while `visit time` contains the time span in which the crawler should visit the site.

Links to pages listed in *robots.txt* can still appear in search results if they are linked to from a page that is crawled. However, the use of an HTML `<meta name="robots" content="noindex" />`[9] in a Web page will suggest that crawlers ignore it.

Note that the protocol is purely advisory, and compliance with the instructions contained in *robots.txt* is more a matter of etiquette than imposition: marking an area of a site out of bounds with robots.txt does not guarantee privacy. If the *robots.txt* file doesn't exist, crawlers might assume that the site owner wishes to provide no specific instructions. A 2007 study [324] analyzed 7,600 websites and found that only 38.5 % of them used *robots.txt*.

The rules listed in *robots.txt* can be ambiguous, or might be differently interpreted by crawlers. A simple way for a webmaster to inform search engines about URLs on a website that are available for crawling is the use of *Sitemaps*.[10] A Sitemap is an XML file (exemplified in Fig. 6.6) that lists URLs for a site along with additional metadata about each resource: when it was last updated, how often the resources usually change, and how important the resource is (compared to others in the site), so that search engines can more intelligently crawl the site.

The Sitemap file is typically located at the root of the URL hierarchy at the site; however, its location can also be included in the robots.txt file by adding to *robots.txt* the directive shown below:

```
Sitemap: http://www.w3.org/sitemap.xml
```

As for the Robots Exclusion Protocol, the use of Sitemaps does not guarantee that Web pages will be included in search indexes. However, major search engines suggest their use, as they will provide a convenient means for the notification of updates in the website resources, thus optimizing the crawling process and, ultimately, reducing the overall indexing time.

6.5 Indexing

Various optimized index structures have been developed to support efficient search and retrieval over text document collections (see Chap. 2), while multimedia col-

---

[9]http://www.robotstxt.org/meta.html.

[10]http://www.sitemaps.org/.

```
<?xml version="1.0" encoding="UTF-8"?>

<urlset xmlns=
     "http://www.sitemaps.org/schemas/sitemap/0.9">
  <url>
     <loc>http://www.webhome.com/</loc>
     <lastmod>2013-01-10</lastmod>
     <changefreq>monthly</changefreq>
     <priority>0.8</priority>
  </url>

  <url>
     <loc>http://www.webhome.com/offers</loc>
     <changefreq>weekly</changefreq>
  </url>

  <url>
     <loc>http://www.webhome.com/news</loc>
     <lastmod>2013-01-22</lastmod>
     <priority>0.3</priority>
  </url>
</urlset>
```

Fig. 6.6 Example Sitemap XML file

lections are typically indexed using data structures optimized for the traversal of high-dimensional descriptors.

Conceptually, building an index involves processing each resource to extract the set of describing features used for storage and retrieval. In the case of textual documents, pages are analyzed to extract index terms, their location in the page, and, ultimately, the associated postings list. After sorting, postings are stored on disk as a collection of inverted lists.

In several applications, where the collection of indexed resources is small and relatively consolidated in time, index construction can be efficiently performed on a single machine. The Web, on the other hand, provides a huge amount of resources, which, by design, can very frequently change during their existence. Therefore, building efficient Web-scale indexes is a critical task, which typically requires the *distribution* and replication of indexes on many servers. Distributing indexes across multiple nodes has several advantages, such as utilizing less computation resources and exploiting data locality [32]: as indexes are smaller, servers can comfortably hold dictionaries in memory, along with the associated housekeeping information, thus saving a disk access for each term during query processing. However, distribution requires a more complex orchestration architecture, and the use of sophisticated optimization techniques to reduce the overall resource utilization (in particular, *caching*).

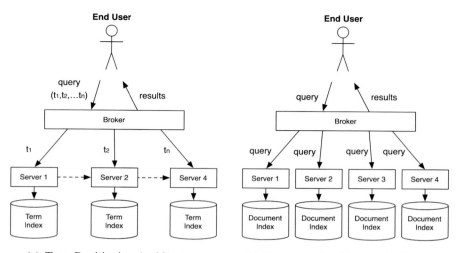

(a) Term Partitioning Architecture   (b) Document Partitioning Architecture

Fig. 6.7 Architectures of (**a**) term partitioning and (**b**) document partitioning approaches for distributed indexes

6.5.1 Distributed Indexing

There are two main strategies for distributing the index of a search engine: by *term partitioning* and by *document partitioning*. The architectures are shown in Fig. 6.7.

In the term partitioning approach, also known as *global inverted files* or *global index organization*, each server hosts a subset of the dictionary of index terms, along with the related postings. Upon query submission, a *broker* node, which is aware of the distribution of terms across servers, analyzes the query terms, asks the appropriate servers to supply the related postings, and combines the inverted lists to generate the final document ranking. Alternative implementations based on pipelined architectures (such as the one proposed in [255]) process queries in an incremental fashion: the broker forwards the query to the first involved term node which, in turn, forwards the query (together with the collected postings list) to the following involved term node, and so on. When the last node in the pipeline is finished, the final result is sent back to the broker, and the top results are returned to the user.

This approach allows reduced per-query overall disk activity [253], as only a subset of the servers are involved in the retrieval, and a greater concurrency, since a stream of queries with different query terms would hit different sets of machines [241]. However, the cost of disk accesses is replaced by network transfers, and the computational bottleneck is moved to the *broker* node: multiterm queries require the transmission of long postings lists between sets of nodes, which then must be merged by the broker. Individual servers do not have enough information to estimate the portion of each inverted list that suffices for query answering, and the broker may be able to estimate this information only by maintaining centralized statistics of the global vocabulary [32]. Finally, with term partitioning, index creation and maintenance are more involved [294], as the definition of partitioning is typically a function of the distribution of query terms and their co-occurrences, which might change with time.

The document partitioning approach, known as *local inverted files*, assumes each node to host complete vocabularies and postings lists for all the locally indexed documents. Upon query submission, the *broker* node broadcasts the query to all nodes, and merges the results from the various nodes before presentation to the user. With respect to term partitioning indexes, less inter-node communication is traded with higher disk activity, and the calculation of a correct global ranking depends on the availability, on each node, of *global* statistics about the entire document collection (e.g., the inverse document frequency (IDF)), which must be refreshed (periodically or upon index variation) by a dedicated background process. On the other hand, document partitioning allows the creation of more robust systems, as query evaluation can proceed even if one of the servers is unavailable, and greatly simplifies the insertion of new documents.

Experimental evaluations [24] have shown that, under the assumption that the query broker receptionist has a large memory and that the distribution of query terms is skewed (thus allowing effective caching), a term-partitioned system can handle greater query throughput than a document-partitioned one. Nonetheless, most large search engines prefer a document-partitioned index (which can be easily generated from a term-partitioned index). Regardless of the approach, query throughput rates can then be resolved by replicating the entire system [36].

6.5.2 Dynamic Indexing

Another distinguishing characteristic of the Web is its continuous evolution. In order to guarantee the availability of updated views on the Web, Web search engines must perform their crawling activities in a continuous fashion, thus requiring a continuous, *dynamic*[11] update of their search indexes. Given that indexing data structures are optimized for efficient retrieval time *and* space occupation, the extension of such structures is a time- and resource-consuming process; thus, dynamic indexing is typically performed *periodically* or *incrementally*.

- *Periodic* reindexing is the simplest approach, and it requires the offline construction of a new index to be used for query processing in place of the old one; after swapping, the old index is deleted. This approach is typically adopted by Web search engines, and it is better suited when the number of changes in the underlying resource collection is relatively small, the latency time for document indexing is acceptable, and the cost associated with the index update is reasonable. Indeed, with periodic indexing, document updates or removals are not an issue, as the document collection is assumed to be consistent at indexing time.

---

[11]*Dynamic* indexing is also referred to as *online* indexing, to emphasize the availability of the system upon document arrival and indexing.

- When the immediate availability of new documents for search is a strict requirement, then the dynamic indexing is performed *incrementally* by maintaining a small in-memory auxiliary index that contains the most recently uploaded documents. Document updates are managed by invalidating old index entries and reindexing the new version; document removal is simply performed by removing deleted documents before result production. Searches are run across the main and the auxiliary indexes, and the results are merged. The size of auxiliary indexes is bound to the amount of available memory. Therefore, when the memory threshold is reached, both indexes are merged into a new on-disk index [100, 223, 329] using ad hoc merging algorithms. While it is efficient for quick document availability, *incremental* indexing complicates the maintenance of collection-wide statistics [241]; for instance, the correct number of hits for a term requires multiple lookups.

6.5.3 Caching

Successful Web search engines must sustain workloads of *billions* of searches per day.[12] To achieve the sub-second response time that users typically expect from search systems, the availability of large-scale infrastructure might not be enough.

Access to primary memory is orders of magnitude faster than access to secondary memory. *Caching* is an effective technique to enhance the performance of a Web search engine. Empirical studies have shown that the stream of queries submitted by users is characterized by a high locality [356], thus allowing caches to improve search responsiveness by retrieving results from the cache without actually querying the search infrastructure.

A search engine can store the results of a query as follows [50, 124, 221]: for each query in the stream of user-submitted search queries, the engine first looks it up in the cache, and if results for that query are stored in the cache—a cache hit—it quickly returns the cached results to the user. Upon a cache miss—when the query results are not cached—the engine evaluates the query and computes its results. The results are returned to the user and are also forwarded to the cache. When the cache is not full, it caches the newly computed results. Otherwise, the cache replacement policy may decide to delete some of the currently cached results to make room for the newly computed set.

Alternatively, the engine can use cache memory to speed up computation by exploiting frequently or recently used evaluation data. For instance, as the search engine evaluates a particular query, it may decide to store in memory the inverted lists of the involved query terms [27, 30, 31]. A large-scale search engine would probably cache both types of information at different levels of its architecture [27]: in [307], it is shown that two-level caching achieves a 52 % higher throughput than

---

[12]http://www.comscore.com/Insights/Press_Releases/2012/11/comScore_Releases_October_2012 _U.S._Search_Engine_Rankings, visited 6th December 2012.

caching only inverted lists, and a 36 % higher throughput than caching only query results.

Independent of the adopted caching policy, an important design decision to address concerns *which* information should be cached. This decision can be taken either *offline* (static caching) or *online* (dynamic caching) [124, 245], but it typically depends on (i) the index update cycles, which might make cache entries stale, thus decreasing the freshness of served results, and (ii) the evaluation of statistics about the search engine usage, possibly produced from the query log history.

In the *offline* case, the cache is filled with data related to the most frequent queries; the cache is read-only, i.e., no replacement policy is applied, and it is periodically updated to reflect changes in the query frequency distribution or changes in the underlying indexes. In the *online* case, the cache content changes dynamically with respect to the query traffic, as new entries may be inserted and existing entries may be evicted. The analysis in [245] reveals that the offline caching strategy performs better when the cache size is small, but online caching becomes more effective when the cache is relatively larger. However, when the cache capacity increases (e.g., several gigabytes) and the problem of frequent index updates is considered, the cache hit rate is mainly dependent on the fraction of unique queries in the query log, and the main scalability problem is the ability to cope with changes to the index [73].

6.6 Exercises

6.1 Describe a *Web miniature* as a graph of interconnected nodes, where nodes represent pages and are associated with a list of words, and arcs represent links between pages. Then, program a crawler whose task is to visit the nodes and to retrieve the list of words from the pages of the Web miniature. Design the crawler's frontier, choose a strategy for visiting nodes, and measure the number of visits which are required to complete the visit of all the nodes. If certain nodes cannot be reached, explain why in terms of the graph's topology.

6.2 Assume that the Web miniature is crawled in parallel; design your version of the *page seen* test.

6.3 Build an index of terms to pages based upon the results produced by crawlers.

Chapter 7
Link Analysis

Abstract Ranking is perhaps the most important feature of a search engine, as it allows the user to efficiently order the huge amount of pages matching a query according to their relevance to the user's information need. With respect to traditional textual search engines, Web information retrieval systems build ranking by combining at least two evidences of relevance: the degree of matching of a page—the content score—and the degree of importance of a page—the popularity score. While the content score can be calculated using one of the information retrieval models described in Chap. 3, the popularity score can be calculated from an analysis of the indexed pages' hyperlink structure using one or more *link analysis* models. In this chapter we introduce the two most famous link analysis models, *PageRank* and *HITS*.

7.1 The Web Graph

The Web's hyperlink structure can be represented as a *directed* graph, where each node represents a Web page and each arc represents a link. The source node of an arc is the Web page containing the link, while the destination node is the linked Web page. The resulting graph is simple: no self-loops are allowed, and even if there are multiple links between two pages only a single edge is placed. Figure 7.1 shows an example of a directed graph representing a Web of six pages.

The graph can be described by an $n \times n$ adjacency matrix $\mathbf{E}$, where for each couple of pages i and j, $\mathbf{E}_{i,j} = 1$ if there is a link from i to j, and $\mathbf{E}_{i,j} = 0$ otherwise. Considering the Web graph in Fig. 7.1, the resulting adjacency matrix is

$$
\mathbf{A} = \begin{array}{c} \\ P_1 \\ P_2 \\ P_3 \\ P_4 \\ P_5 \\ P_6 \end{array}
\begin{array}{c} \begin{array}{cccccc} P_1 & P_2 & P_3 & P_4 & P_5 & P_6 \end{array} \\
\left(\begin{array}{cccccc}
0 & 0 & 0 & 0 & 1 & 0 \\
0 & 0 & 0 & 1 & 1 & 0 \\
1 & 0 & 0 & 0 & 0 & 0 \\
1 & 0 & 0 & 0 & 0 & 1 \\
0 & 0 & 1 & 0 & 0 & 0 \\
1 & 0 & 0 & 0 & 0 & 0
\end{array} \right) \end{array}
\tag{7.1}
$$

S. Ceri et al., *Web Information Retrieval*, Data-Centric Systems and Applications,
DOI 10.1007/978-3-642-39314-3_7, © Springer-Verlag Berlin Heidelberg 2013

Fig. 7.1 An example of a
page graph composed of 6
pages and 8 links

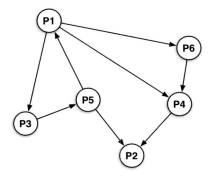

Hyperlinks directed into a page are referred as *inlinks*, whole those originating
from a page are called *outlinks*. The *indegree* of a page i is calculated as the number
of edges coming *into* i; likewise, the *outdegree* of node i is defined as the number
of edges coming *out* from i. In the example of Fig. 7.1, page 1 has an *indegree* of
1 and an *outdegree* of 3. A directed graph is weakly connected if any node can be
reached from any other node by traversing edges either in their indicated direction
or in the opposite direction.

The Web graph has been intensively studied over recent years, and it has been
empirically shown to be far from strongly connected (intuitively, not all the pages
in the graph are mutually linked). Many works addressed network properties of
the Web, showing a scale-free nature of the link distributions, with small-world
properties[1] and preferential attachment [14]: the way new nodes link with other
nodes is not driven by causality, but strongly depends on the node degrees—the
more connected a node is, the more likely it is to receive new links. Theoretical
models have shown that, in the large, the Web graph can be modeled in terms of
degree distribution which follows a *power law*:

$$\Pr(outdegree = k)\frac{1}{k^{a_{\text{out}}}}$$
$$\Pr(indegree = k)\frac{1}{k^{a_{\text{in}}}} \tag{7.2}$$

i.e., the probability that the outdegree (indegree) is equal to k is inversely proportional to k^a, for some model parameter a. The Barabasi–Albert model applied to the
Web graph [14] estimated a value of $a = 3$, and a diameter of 19; i.e., two randomly
chosen documents on the Web are on average 19 clicks away from each other. Measures on the actual Web graph reported values of a ranging from exponents between
2.1 and 2.5. The number of inlinks to a page has been empirically evaluated to be,
on average, from roughly 8 to 15 [241].

In a seminal study [67], authors mapped a large Web crawl of over 200 million
nodes and 1.5 billion hyperlinks, showing that, unlike what was estimated by previous studies, the Web is not structured as clusters of sites forming a well-connected

---

[1]The diameter of the graph is logarithmically proportional to the number of its nodes.

Fig. 7.2   The bow-tie
structure of the Web [67]

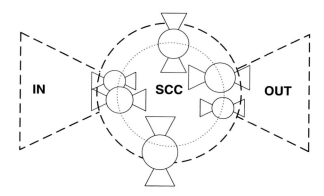

sphere. Instead, the Web graph resembles a bow-tie shape, as illustrated in Fig. 7.2, consisting of three major regions (a knot and two bows), and a fourth, smaller region of pages that are "disconnected" from the basic bow-tie structure.

The bow-tie model identifies a *core*, accounting for around 30 % of the pages, which is made by large strongly connected components (SCCs), i.e., pages connected by extensive cross-linking among and between themselves that are more likely to be visited by users and search engine crawlers. The left *IN bow* includes around 24 % of the pages, and it contains upstream nodes, i.e., pages that allow users to reach the core but that have not yet been linked to; these pages are typically new or unpopular Web pages that haven't yet attracted interest from the Web community. The right *OUT bow* also includes around 24 % of the pages, and contains downstream nodes, pages that can be reached from the core but do not link to it (for instance, corporate or commercial websites that contains only internal links). A fourth group of pages, called *tendrils*, contain nodes that are linked by an upstream or downstream node, but can neither reach nor be reached from the core. Finally, around 22 % of the pages form *disconnected components*, i.e., small groups of nodes not connected to the bow tie.

In the same paper, the researchers also confirmed that many features of the Web graph follow a power law distribution, but with different parameters. For instance, they showed that if a direct hypertext path does exist between randomly chosen pages, its average length is 16 links. Moreover, for any randomly chosen source and destination page, the probability that a direct hyperlink path exists from the source to the destination is only 24 %.

7.2 Link-Based Ranking

In addition to being the structural backbone of the simple yet successful user navigation paradigm of the Web, hyperlinks can also be viewed as evidences of endorsement, or *recommendation*; under this assumption, a page with more recommendations (incoming links) can be considered more important than a page with less incoming links. However, for a recommendation to be relevant, the recommender should possess some sort of *authority*. Likewise, for a recommender to be authoritative, it must be considered a *relevant* source of recommendation.

According to the base set of pages considered for the analysis, we can distinguish ranking systems into query-independent and query-dependent systems.

- *Query-independent systems* aim at ranking the entire Web, typically only based on the authoritativeness of the considered results. Given the size of the problem, authoritativeness is computed offline, but it is considered at query time as an additional ranking component. The most popular example of query-independent ranking systems is *PageRank*, which is described in Sect. 7.3.
- *Query-dependent systems* work on a reduced set of resources, typically the ones that have been identified as related to the submitted query. They can be computed online at query time, and they provide an additional measure of recommendation (hub) score. The most popular example of query-dependent ranking systems is *HITS*, which is described in Sect. 7.4.

A simple heuristic for the calculation of the popularity score of pages is to rank them according to their visibility, i.e., by the number of pages that link to them. The intuition is that a good authority is a page that is pointed to by many nodes in the Web graph; therefore, the popularity score of a page P_i can be calculated by its *indegree* $\|B_{P_i}\|$, where B_{P_i} are the pages linking into P_i. This simple heuristic was applied by several search engines in the early days of Web search [51]. However, several works [203] recognized that this approach was not sophisticated enough to capture the authoritativeness of a node, as it suffers from a major drawback: it considers all links to a page to be equally important. Thus, the most important link-based ranking methods, discussed in the next two sections, start from the assumption that a link to page *j* from an important page should boost the page *j* importance score more than a link from an unimportant one. However, the *indegree* score allows us to achieve good ranking performance, as discussed in Sect. 7.5.

7.3 PageRank

In 1995 Sergey Brin and Larry Page were two computer science students at Stanford University when they started to collaborate on their Web search engine, Google. In August 2008 they took a leave of absence from Stanford in order to focus on their creature; in the same year, they presented a milestone publication at the Seventh International World Wide Web Conference (WWW 1998): "The anatomy of a large-scale hypertextual Web search engine" [64]. While setting the research agenda for the decades to come, the paper (and following publications [271]) provided a first introduction to the *PageRank* link analysis method adopted (at the time) in Google.

The *PageRank* method focuses only on the *authority* dimension of a page and, in its simpler version, considers a *Web page important if it is pointed to by another important page*, thus assigning a single *popularity* score to it, the *PageRank score*.

Fig. 7.3 Directed graph
representation of a small Web
graph

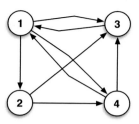

This score is *query independent*, that is, its value is determined offline and it remains
constant regardless of the query.

Let us consider the small Web graph of Fig. 7.3: we can model the list of page
scores associated with each node as a variable x, and represent the nodes in the
graph as the vector $\mathbf{x}$ in Eq. (7.3):

$$\mathbf{x} = [x_1 \quad x_2 \quad x_3 \quad x_4]^T \tag{7.3}$$

The PageRank score of a page x_i can be calculated as the sum of the scores of
the nodes linking *to* x_i normalized by the number of their outgoing links. More
formally:

$$x_i = \sum_{j \in L_i} \frac{x_j}{n_j} \tag{7.4}$$

where L_i is the set of x_i's backlinks, and n_j is the number of x_j's outgoing links.

The value of the PageRank scores for the nodes in the graph can thus be calculated by solving the resulting system of linear equations. For instance, considering
the graph in Fig. 7.3:

$$\begin{aligned}
x_1 &= \frac{x_3}{1} + \frac{x_4}{2} \\
x_2 &= \frac{x_3}{1} \\
x_3 &= \frac{x_1}{3} + \frac{x_2}{2} + \frac{x_4}{2} \\
x_4 &= \frac{x_1}{3} + \frac{x_2}{2}
\end{aligned} \tag{7.5}$$

Solving the system in Eq. (7.5) is equivalent to solving the system

$$\mathbf{A}\,\mathbf{x} = \mathbf{x} \tag{7.6}$$

where $\mathbf{x}$ is an $n \times 1$ PageRank scores vector, and $\mathbf{A}$ is the $n \times n$ weighted adjacency matrix of the n pages' graph; note that $\mathbf{A}_{ij} = 0$ if there is no link between the
nodes i and j, while $\mathbf{A}_{ij} = \frac{1}{n_j}$ otherwise. Considering the Web graph in Fig. 7.3,

the resulting system is

$$
\begin{bmatrix} 0 & 0 & 1 & \frac{1}{2} \\ \frac{1}{3} & 0 & 0 & 0 \\ \frac{1}{3} & \frac{1}{2} & 0 & \frac{1}{2} \\ \frac{1}{3} & \frac{1}{2} & 0 & 0 \end{bmatrix} \begin{bmatrix} x_1 \\ x_2 \\ x_3 \\ x_4 \end{bmatrix} = \begin{bmatrix} x_1 \\ x_2 \\ x_3 \\ x_4 \end{bmatrix}
\tag{7.7}
$$

Recalling our knowledge of linear algebra, we know that solving the system in Eq. (7.7) means finding one *eigenvector* corresponding to the *eigenvalue* 1. Since PageRank reflects the relative importance of the nodes, we chose as PageRank vector the eigenvector for which the sums of all entries are equal to 1.

7.3.1 Random Surfer Interpretation

The weights assigned to the edges on the graph can be interpreted in a probabilistic way, according to the *random surfer* abstractions introduced by Page and Brin, where the PageRank score associated with a page i is the probability that a surfer randomly exploring the Web graph will visit such a page.

As the importance of a Web page can be associated with the number of incoming links it has, $\mathbf{A}_{ij}$ can be viewed as the *probability* of going from page j to page i, and the process of finding the PageRank vector as a *random walk* on the Web graph. For instance, in the example of Fig. 7.3, a random surfer currently visiting the page 1 has $\frac{1}{3}$ probability to go to page 3, 4, or 5.

Given an initial PageRank vector $\mathbf{x}_0$ where all pages have equal probability to be visited, the probability that a page i will be visited after one navigation step is equal to $\mathbf{A}\mathbf{x}_0$, while the probability that the same page will be visited after k steps is equal to $\mathbf{A}^k \mathbf{x}_0$.

The iterative process described above is known as the *power iteration method*, a simple iterative method for finding the dominant eigenvalue and eigenvector of a matrix, and it can be used to obtain the final PageRank vector.

$$
\mathbf{x}^* = \lim_{k \to \infty} \mathbf{A}^k \mathbf{x}_0
\tag{7.8}
$$

The power method in Eq. (7.8) is a linear stationary process [215], where $\mathbf{A}$ can be seen as a *stochastic transition probability matrix* for a Markov chain.

If the considered Web graph is strongly connected (i.e., there is a path from each vertex in the graph to every other vertex), the sequence $\mathbf{A}^1 \mathbf{x}, \mathbf{A}^2 \mathbf{x}, \ldots, \mathbf{A}^k \mathbf{x}$ is guaranteed to *converge* to a unique probabilistic vector $\mathbf{x}^*$, the final PageRank vector.

Let us consider the strongly connected Web graph of Fig. 7.3. Assuming an initial PageRank vector $\mathbf{x}_0 = [0.25\ 0.25\ 0.25\ 0.25]^T$, the scores for the pages in the graph evolve as depicted in Fig. 7.4, where the final PageRank vector converges to

$$
\mathbf{x}^* = [0.387 \quad 0.129 \quad 0.290 \quad 0.194]^T
\tag{7.9}
$$

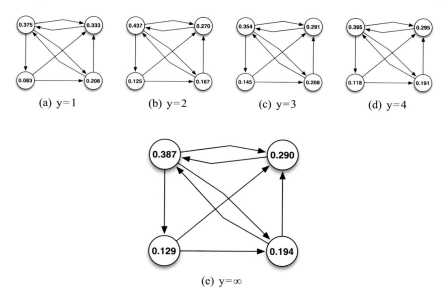

Fig. 7.4 Power iterations at (**a**) $k = 1$, (**b**) $k = 2$, (**c**) $k = 3$, (**d**) $k = 4$, and (**e**) $k = \infty$ for the PageRank method applied to the Web graph of Fig. 7.3

The component x_i^* of the PageRank vector represents the *popularity* of the Web page i, viewed as the *probability* that a Web surfer following a *random walk* will arrive on the page i at some point in the future. Equivalently, one can interpret the PageRank value $\mathbf{x}_i$ as the fraction of time that the surfer spends, in the long run, on page i: intuitively, pages that are often visited during the random walk must be important.

The PageRank scores can be used to rank pages by their importance. In our example, page 1 is the most important page, followed by page 2, page 3, and page 4. Recent works [39, 272] have shown how, due to the scale-free nature of the Web, the values of the PageRank scores follow a long-tail distribution;[2] that is, very few pages have a significant PageRank while the vast majority of pages have a very low PageRank value.

Strong connectivity is almost impossible to reach in the Web graph, mainly because of two typical phenomena: *dangling nodes* and disconnected subgraphs.

7.3.2 Managing Dangling Nodes

Dangling nodes are nodes in the graph which have no outgoing links to other nodes. Concrete examples of dangling nodes are non-HyperText Markup Language (HTML) resources such as PDF files, images, and videos.

---

[2]More precisely, a power law distribution with $a \approx 2.1$.

Fig. 7.5 Directed graph
representation of a small Web
graph with dangling nodes

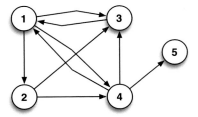

Due to the presence of dangling nodes, the weighted adjacency matrix $\mathbf{A}$ might not be *column-stochastic*; i.e., there might be columns for which the sum of the entries is not 1. As the stochasticity property is missing, the matrix $\mathbf{A}$ is not a Markov chain's *transition probability matrix*, and, therefore, it does not have an eigenvalue equal to 1.

If we consider the Web graph example of Fig. 7.5, which corresponds to Eq. (7.10), node 5 is a dangling node, and the corresponding column in the matrix $'$ makes this matrix not column-stochastic.

$$
\begin{bmatrix}
0 & 0 & 1 & \frac{1}{3} & 0 \\
\frac{1}{3} & 0 & 0 & 0 & 0 \\
\frac{1}{3} & \frac{1}{2} & 0 & \frac{1}{3} & 0 \\
\frac{1}{3} & \frac{1}{2} & 0 & 0 & 0 \\
0 & 0 & 0 & \frac{1}{3} & 0
\end{bmatrix}
\begin{bmatrix}
x_1 \\ x_2 \\ x_3 \\ x_4 \\ x_5
\end{bmatrix}
=
\begin{bmatrix}
x_1 \\ x_2 \\ x_3 \\ x_4 \\ x_5
\end{bmatrix}
\tag{7.10}
$$

To cope with the problem, two solutions are commonly available. The first solution simply requires the *removal of dangling nodes*, performed *before* the construction of $\mathbf{A}$, which then becomes stochastic. The PageRank scores are then calculated using the power iteration methods, and PageRank scores for the removed nodes are computed based on *incoming link* only, e.g., by using a weighted average of the PageRank of the incoming source nodes of the incoming links. Considering the example of Fig. 7.5, by removing node 5 the PageRank vector of the reduced matrix is calculated as in Eq. (7.11). The final PageRank vector is calculated by adding the score for the dangling nodes; in our example, we decide $x_5 = \frac{x_4}{2} = 0.097$, and the final PageRank vector is

$$
\mathbf{x}^* = [0.387 \quad 0.129 \quad 0.290 \quad 0.194 \quad 0.097]^T
\tag{7.11}
$$

The second solution, applied by Brin and Page, involves the definition of a *stochastic adjustment* to apply to all the $\mathbf{0}$ columns of $\mathbf{A}$: specifically, all the elements of such columns are replaced with $\frac{1}{n}$, to represent the possibility for the random surfer to exit a dangling node by hyperlinking to any page at random. We call the resulting matrix $\mathbf{S}$, which is the result of a *rank-1 update* of $\mathbf{A}$:

$$
\mathbf{S} = \mathbf{A} + \mathbf{a}\left(\frac{1}{n}\right)\mathbf{e}^T
\tag{7.12}
$$

Fig. 7.6 An example of a disconnected Web graph

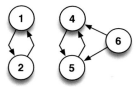

where $\mathbf{e}$ is a vector of all 1, and $\mathbf{a}$ is the *dangling node* vector, where $a_i = 1$ if the node i represents a dangling node, or 0 otherwise. For the example of Fig. 7.5, the matrix becomes

$$
\begin{bmatrix}
0 & 0 & 1 & \frac{1}{3} & 0 \\
\frac{1}{3} & 0 & 0 & 0 & 0 \\
\frac{1}{3} & \frac{1}{2} & 0 & \frac{1}{3} & 0 \\
\frac{1}{3} & \frac{1}{2} & 0 & 0 & 0 \\
0 & 0 & 0 & \frac{1}{3} & 0
\end{bmatrix}
+ \frac{1}{5}
\begin{bmatrix} 0 & 0 & 0 & 0 & 1 \end{bmatrix}
\begin{bmatrix} 1 \\ 1 \\ 1 \\ 1 \\ 1 \end{bmatrix}
=
\begin{bmatrix}
0 & 0 & 1 & \frac{1}{3} & \frac{1}{5} \\
\frac{1}{3} & 0 & 0 & 0 & \frac{1}{5} \\
\frac{1}{3} & \frac{1}{2} & 0 & \frac{1}{3} & \frac{1}{5} \\
\frac{1}{3} & \frac{1}{2} & 0 & 0 & \frac{1}{5} \\
0 & 0 & 0 & \frac{1}{3} & \frac{1}{5}
\end{bmatrix}
\tag{7.13}
$$

The applied adjustment guarantees that $\mathbf{S}$ is stochastic [215, 320], and thus a transition probability matrix for a Markov chain; the PageRank equation then becomes

$$
\mathbf{x}^* = \lim_{k \to \infty} \mathbf{S}^k \mathbf{x}_0
\tag{7.14}
$$

7.3.3 Managing Disconnected Graphs

Due to the presence of disconnected subgraphs, power iterations may not always converge to a *unique* solution. Let's take as an example the Web graph of Fig. 7.6.

The Web graph does not contain dangling nodes, but it clearly shows disconnected subgraphs. The resulting matrix $\mathbf{S}$ is

$$
\begin{bmatrix}
0 & 1 & 0 & 0 & 0 \\
1 & 0 & 0 & 0 & 0 \\
0 & 0 & 0 & 1 & \frac{1}{2} \\
0 & 0 & 1 & 0 & \frac{1}{2} \\
0 & 0 & 0 & 0 & 0
\end{bmatrix}
\tag{7.15}
$$

As $\mathbf{S}$ is not *primitive*, the equation $\mathbf{x} = \mathbf{A}\mathbf{x}$ is satisfied by both vectors $\mathbf{x}^1 = [\frac{1}{2} \ \frac{1}{2} \ 0 \ 0 \ 0]^T$ and $\mathbf{x}^2 = [0 \ 0 \ \frac{1}{2} \ \frac{1}{2} \ 0]^T$, and by any linear combination of $\mathbf{x}^1$ and $\mathbf{x}^2$. To cope with this problem, Brin and Page introduced a second adjustment to guarantee the *primitivity* of the matrix. They built on the same random surfer abstraction by introducing the idea of *teleportation*, i.e., a periodically activated "jump" from the current page to a random one, which is represented by the following *Google matrix* $\mathbf{M}$:

$$
\mathbf{M} = \alpha \mathbf{S} + (1 - \alpha) \mathbf{E}
\tag{7.16}
$$

where α is the *damping factor*, a parameter (a scalar between 0 and 1) that controls the proportion of the time that the random surfer follows the hyperlink structure of the Web graph as opposed to teleporting: when the surfer is currently at a page i with outlinks, it will randomly choose one of these links with probability α, while it will jump to any randomly selected page on the Web with probability $1 - \alpha$. $\mathbf{E}$ is the *teleportation matrix* that uniformly guarantees to all nodes an equal likelihood of being the destination of a teleportation, and it is defined as

$$\mathbf{E} = \frac{1}{n}\mathbf{ee}^T \tag{7.17}$$

In [64] Page and Brin reported the usage of a $\alpha = 0.85$ value. Considering the Web graph of Fig. 7.6, and setting $\alpha = 0.85$, the new matrix $\mathbf{M}$ is

$$\begin{bmatrix} 0.03 & 0.88 & 0.03 & 0.03 & 0.03 \\ 0.88 & 0.03 & 0.03 & 0.03 & 0.03 \\ 0.03 & 0.03 & 0.03 & 0.88 & 0.455 \\ 0.03 & 0.03 & 0.88 & 0.03 & 0.455 \\ 0.03 & 0.03 & 0.03 & 0.03 & 0.03 \end{bmatrix} \tag{7.18}$$

The PageRank score is now computed as

$$\mathbf{x}^* = \lim_{k \to \infty} \mathbf{M}^k \mathbf{x}_0 \tag{7.19}$$

In the example of Fig. 7.6, the unique Google PageRank vector is given by $\mathbf{x}^* = [0.2\ 0.2\ 0.285\ 0.285\ 0.03]^T$.

7.3.4 Efficient Computation of the PageRank Vector

The power method discussed in the previous section is not the only available tool for the computation of the PageRank vector, which we recall being the dominant eigenvector of $\mathbf{S}$ corresponding to the dominant eigenvalue 1 (as $\mathbf{S}$ is stochastic). Also, note that $\mathbf{S}$ is an $n \times n$ positive matrix with no nonzero elements, thus making the calculation of $\mathbf{S}^k\mathbf{x}$ unfeasible for big values of n.

However, the application of the power methods to the Google matrix allows some interesting optimizations: $\mathbf{M}$ can be rewritten as a rank-1 update to the (very sparse) weighted adjacency matrix $\mathbf{A}$, as shown in Eq. (7.20).

$$\begin{aligned} M &= \alpha\mathbf{S} + (1-\alpha)\frac{1}{n}\mathbf{ee}^T \\ &= \alpha\left(\mathbf{A} + \frac{1}{n}\mathbf{ae}^T\right) + (1-\alpha)\frac{1}{n}\mathbf{ee}^T \\ &= \alpha\mathbf{A} + \left(\alpha\mathbf{a} + (1-\alpha)\mathbf{e}\right)\frac{1}{n}\mathbf{e}^T \end{aligned} \tag{7.20}$$

$\mathbf{A}$ is a very sparse matrix which never changes during computation, while the dangling node vector $\mathbf{a}$ and the identity vector $\mathbf{e}$ are constant. Therefore, $\mathbf{x}^{k+1}$ can be very efficiently computed only through matrix–vector multiplications, and only by storing the $\mathbf{A}$, $\mathbf{a}$, and the current PageRank vector $\mathbf{x}^k$.

The power method was also chosen by Page and Brin, as it has been empirically shown to require few iterations to converge: in [64] the authors reported that only 50–100 power iterations are needed to converge to a satisfactory PageRank vector. Studies have shown that α has a profound influence on the performance of the PageRank calculation, as it affects the number of required iterations and the sensitivity of the vector values [215].

7.3.5 Use of PageRank in Google

As repeatedly stated by several Google employees, PageRank was a very important component of the Google Web search engine in the first years of its existence. However, PageRank is now only one of the many factors that determine the final score, and ultimately, the result ranking, of a Web page in Google.

While it is still being used,[3] PageRank is now a part of a much larger ranking system that it is said to account for more than 200 different "signals" (ranking variables). These include (i) *language* features (phrases, synonyms, spelling mistakes, etc.), (ii) *query* features that relate to language features, trending terms/phrases, and *time*-related features (e.g., "news" related queries might be best answered by recently indexed documents, while factual queries are better answered by more "resilient" pages); and (iii) *personalization* features, which relate to one's search history, behavior, and social surrounding.

However, Google still very carefully guards the exact PageRank of pages, while offering to the public an approximate value (through its Google Toolbar—Fig. 7.7) measured on a logarithmic scale of one to ten. This value is given just as a rough indication of the overall popularity of a Web page, but several studies (conducted mainly by search engine optimization companies), have shown that its value its very loosely correlated with the actual placement in the result list of the related page. Also, the score displayed in the Google Toolbar is updated only a few times a year,[4] and it does not reflect the actual value fluctuations due to changes in Google's ranking algorithms.

7.4 Hypertext-Induced Topic Search (HITS)

Developed in IBM by Jon Kleinberg in 1998 [203], the *Hypertext-Induced Topic Search* (HITS) link analysis model (also known as Hyperlink-Induced Topic Search)

---

[3]http://googleblog.blogspot.it/2008/05/introduction-to-google-search-quality.html.

[4]http://googlewebmastercentral.blogspot.it/2011/06/beyond-pagerank-graduating-to.html.

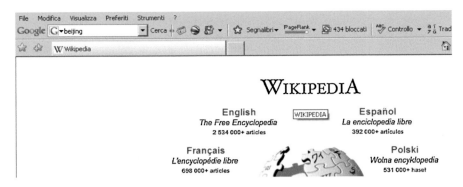

Fig. 7.7 A screenshot of the Google PageRank value for the home page of Wikipedia.org as shown by Google Toolbar

is one of the pioneering methods for the exploitation of the Web hyperlink structure, and it is currently adopted by several search services including Taoma[5] and Ask.com.[6]

Although very similar to *PageRank*, *HITS* exploits both the inlinks and outlinks of Web pages P_i to create two *popularity* scores: (i) the *hub* score, $h(P_i)$, and (ii) the *authority* score, $a(P_i)$.

The *hub* score provides an indication of the importance of the links which exit the node and is used in order to select the pages containing relevant information, whereas the *authority* score measures the value of the links which enter the node and is used to describe contents.

Starting from this informal definition of *hub* and *authority* scores, we can identify as good *hubs* those pages with many outlinks that points to good authorities; for instance, directories like *Yahoo! Directory* (http://dir.yahoo.com/) contain links to high-quality, manually reviewed resources. Likewise, good *authorities* are those pages with many inlinks coming from good hubs; for instance, pages in *Wikipedia* are likely to be considered good authorities, as they are more likely to be linked to topic-specific pages.

The implementation of HITS involves two main steps: (i) the construction of the Web graph stemming from the submitted query, and (ii) the calculation of the final authority and hub scores.

7.4.1 Building the Query-Induced Neighborhood Graph

Unlike PageRank, HITS is *query dependent*, as the link analysis is carried out at query time over a query-induced graph.

---

[5]http://www.taoma.com.

[6]http://www.ask.com.

Fig. 7.8 Example of
neighborhood graph
expansion in HITS

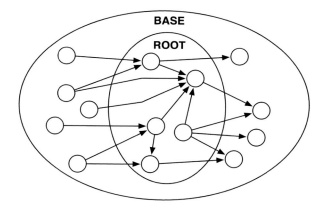

When the query q is sent to the search engine, the system returns a *root* set of
pages, possibly using one of the retrieval models defined in Chap. 3. This graph is
expanded by adding pages that either point to (inlinks) or are pointed by (outlinks)
pages in the root set, so as to create a *base* set of pages, as exemplified in Fig. 7.8.

This expansion set exploits the locality of contents on the Web to increase the
overall *recall* of the user search by enacting a *topic*-based expansion of the original
query results: intuitively, given a page about a topic (e.g., "computer"), by following
an outgoing link one is more likely to find another page on the same topic (e.g.,
"laptop") than by random sampling a page from the Web.

However, it has been empirically shown that small additions to a neighborhood
graph may considerably change the authority and hub scores for pages. Therefore,
the radius of expansion is typically limited to pages at distance one from the *root* set
of pages, and, in practice, the number of considered inlinks and outlinks is bounded
to a given threshold, e.g., 100. More details about the sensitivity of HITS to variations of the neighborhood graph will be discussed in Sect. 7.4.4.

7.4.2 Computing the Hub and Authority Scores

As in PageRank, the values of hub and authority scores for pages in the graph are
calculated in an iterative fashion.

The authority score $\mathbf{a}(P_j)$ of a page j is proportional to the sum of the hub scores
$\mathbf{h}(P_j)$ of the pages $P_j \in BI_{P_i}$, where BI_{P_i} are the pages linking into P_i; conversely,
the hub score $\mathbf{h}(P_i)$ of a page i is proportional to the authority scores $\mathbf{a}(P_j)$ of the
pages $P_j \in BO_{P_i}$, where BO_{P_i} are the pages to which P_i links.

Let $\mathbf{E}$ denote the adjacency matrix of the directed graph composed by the *base*
set of pages, where $\mathbf{E}_{ij} = 1$ if there is a link from the node i to the node j, while
$\mathbf{E}_{ij} = 1$ otherwise.

Fig. 7.9 Directed graph of a
base set of pages resulting
from a query q

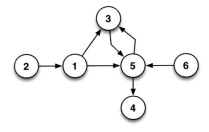

The hub and authority scores can be calculated by iteratively solving the following equation systems:

$$\tilde{\mathbf{a}}^k = \mathbf{E}^T \mathbf{h}^{k-1} \quad \mathbf{a}^k = \frac{\tilde{\mathbf{a}}^k}{\|\tilde{\mathbf{a}}^k\|_1} \tag{7.21}$$

$$\tilde{\mathbf{h}}^k = \mathbf{E}\mathbf{a}^k \quad \mathbf{h}^k = \frac{\tilde{\mathbf{h}}^k}{\|\tilde{\mathbf{h}}^k\|_1} \tag{7.22}$$

where $\mathbf{a}^k$ and $\mathbf{h}^k$ are vectors that hold the authority and hub scores at the current
iteration, and $\tilde{\mathbf{a}}^k$ and $\tilde{\mathbf{h}}^k$ are their normalized versions. The computation stops when
the vectors $\tilde{\mathbf{a}}^k$ and $\tilde{\mathbf{h}}^k$ converge to a final solution.

From Eqs. (7.21) and (7.22) is it possible to intuitively understand the idea behind the HITS method. Top hub pages increase the authority score of pages they
point to (Eq. (7.21)), i.e., top hubs "vote for" top authorities; on the other hand, top
authority pages increase the hub score of pages they point to (Eq. (7.22)), that is, top
authorities "vote for" top hubs. Upon convergence, vectors $\mathbf{a}$ and $\mathbf{h}$ can be sorted to
report top-rank authority and top-rank hub pages.

To exemplify the adoption of the HITS algorithm to a Web graph, let us consider
the example of Fig. 7.9, for which the associated adjacency matrix $\mathbf{E}$ is

$$\mathbf{E} = \begin{bmatrix} 0 & 0 & 1 & 0 & 1 & 0 \\ 1 & 0 & 0 & 0 & 0 & 0 \\ 0 & 0 & 0 & 0 & 1 & 0 \\ 0 & 0 & 0 & 0 & 0 & 0 \\ 0 & 0 & 1 & 1 & 0 & 0 \\ 0 & 0 & 0 & 0 & 1 & 0 \end{bmatrix} \tag{7.23}$$

Assuming an initial hub score vector $\mathbf{h}_0 = [0.17\ 0.17\ 0.17\ 0.17]^T$, the scores
for the pages in the graph evolve as depicted in Fig. 7.10, where the final hub and
authority scores vectors converge to:

$$\begin{aligned} \mathbf{a}^* &= [0 \quad 0 \quad 0.37 \quad 0.13 \quad 0.50 \quad 0]^T \\ \mathbf{h}^* &= [0.36 \quad 0 \quad 0.21 \quad 0 \quad 0.21 \quad 0.21]^T \end{aligned} \tag{7.24}$$

meaning that page 3 is the most authoritative, while page 1 is the best hub for the
query.

Notice that the order of the matrix $\mathbf{E}$ is significantly smaller than the total number of pages on the Web; therefore, the cost of computing the authority and hub

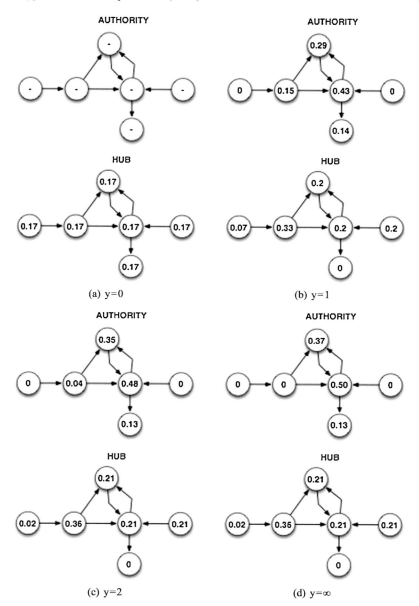

Fig. 7.10 Power iterations at (**a**) $k = 0$, (**b**) $k = 1$, (**c**) $k = 2$, and (**d**) $k = \infty$ for the HITS method applied to the Web graph of Fig. 7.9

scores is small and suitable for online processing. Typically, runs with several thousand nodes and links converge in about 20–30 iterations [85], while several studies

Fig. 7.11 Directed graph of a set of base pages for which several authority and hub vectors exist

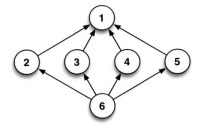

have shown that 10–15 iterations are required for typical (smaller) query-induced graphs.

Similarly to PageRank, the computation of **a** and **h** can be cast as the problem of seeking an eigenvector associated to the largest eigenvalue of a square matrix. The two iterative equations, (7.21) and (7.22), can be simplified by substitution as:

$$\tilde{\mathbf{a}}^k = \mathbf{E}^T \mathbf{E} \mathbf{a}^{k-1} \quad \mathbf{a}^k = \frac{\tilde{\mathbf{a}}^k}{\|\tilde{\mathbf{a}}^k\|_1} \tag{7.25}$$

$$\tilde{\mathbf{h}}^k = \mathbf{E} \mathbf{E}^T \mathbf{h}^{k-1} \quad \mathbf{h}^k = \frac{\tilde{\mathbf{h}}^k}{\|\tilde{\mathbf{h}}^k\|_1} \tag{7.26}$$

The substitution of $\mathbf{E}^T \mathbf{h}^{k-1}$ and $\mathbf{E} \mathbf{a}^k$ with, respectively, $\mathbf{E}^T \mathbf{E} \mathbf{a}^{k-1}$ and $\mathbf{E} \mathbf{E}^T \mathbf{h}^{k-1}$ allows the separation of the score equations, which are now independent. The hub score converges to an eigenvector corresponding to the largest eigenvalue of $\mathbf{E} \mathbf{E}^T$, while the authority score converges to an eigenvector corresponding to the largest eigenvalue of $\mathbf{E}^T \mathbf{E}$. $\mathbf{A} = \mathbf{E}^T \mathbf{E}$ and $\mathbf{H} = \mathbf{E} \mathbf{E}^T$ are now called, respectively, the *authority matrix* and the *hub matrix*. The power iteration method can then be used to solve the system of equations iteratively, initializing the hub vector to a given value.

Considering the example of Fig. 7.9, the corresponding *authority* and *hub* matrices are shown below, and the normalized authority and hub scores for the example are the same as in Eq. (7.24).

$$\mathbf{E} \mathbf{E}^T = \begin{bmatrix} 2 & 0 & 1 & 0 & 1 & 1 \\ 0 & 1 & 0 & 0 & 0 & 0 \\ 1 & 0 & 1 & 0 & 0 & 1 \\ 0 & 0 & 0 & 0 & 0 & 0 \\ 1 & 0 & 0 & 0 & 2 & 0 \\ 1 & 0 & 1 & 0 & 0 & 1 \end{bmatrix} \tag{7.27}$$

$$\mathbf{E} \mathbf{E}^T = \begin{bmatrix} 1 & 0 & 0 & 0 & 0 & 0 \\ 0 & 0 & 0 & 0 & 0 & 0 \\ 0 & 0 & 2 & 1 & 1 & 0 \\ 0 & 0 & 1 & 1 & 0 & 0 \\ 0 & 0 & 1 & 0 & 3 & 0 \\ 1 & 0 & 1 & 0 & 0 & 1 \end{bmatrix} \tag{7.28}$$

7.4.3 Uniqueness of Hub and Authority Scores

Although convergence has been empirically shown to be quickly achievable, there are cases when the uniqueness of the authority and hub scores cannot be guaranteed. One such case is presented in the example of Fig. 7.11, corresponding to the adjacency matrix of Eq. (7.29):

$$\mathbf{E} = \begin{bmatrix} 0 & 0 & 0 & 0 & 0 & 0 \\ 1 & 0 & 0 & 0 & 0 & 0 \\ 1 & 0 & 0 & 0 & 0 & 0 \\ 1 & 0 & 0 & 0 & 0 & 0 \\ 1 & 0 & 0 & 0 & 0 & 0 \\ 0 & 1 & 1 & 1 & 1 & 0 \end{bmatrix} \tag{7.29}$$

The matrix shows the presence of both a row and column of zeros and this causes the authority and hub matrices to become *reducible*, causing Eqs. (7.25) and (7.26) to be satisfied by several vectors (and linear combinations of them). In the example, the equations are satisfied by both vectors:

$$\begin{aligned} \mathbf{a}_1 &= [1 \quad 0 \quad 0 \quad 0 \quad 0 \quad 0]^T \\ \mathbf{a}_2 &= [0 \quad 0.25 \quad 0.25 \quad 0.25 \quad 0.25 \quad 0]^T \end{aligned} \tag{7.30}$$

This problem is also common to PageRank, and can be solved by the *teleportation* matrix, which was used to overcome the *reducibility* problem. In the same spirit, all the entries of $\mathbf{E}$ could be added with a constant value such that $\mathbf{A}$ and $\mathbf{H}$ become irreducible. Considering the example of Fig. 7.11, $\mathbf{E}$ becomes

$$\mathbf{E} = \begin{bmatrix} 0.1 & 0.1 & 0.1 & 0.1 & 0.1 & 0.1 \\ 1.1 & 0.1 & 0.1 & 0.1 & 0.1 & 0.1 \\ 1.1 & 0.1 & 0.1 & 0.1 & 0.1 & 0.1 \\ 1.1 & 0.1 & 0.1 & 0.1 & 0.1 & 0.1 \\ 1.1 & 0.1 & 0.1 & 0.1 & 0.1 & 0.1 \\ 0.1 & 1.1 & 1.1 & 1.1 & 1.1 & 0.1 \end{bmatrix} \tag{7.31}$$

which produces unique hub and authority scores:

$$\begin{aligned} \mathbf{a}_1 &= [0.30 \quad 0.16 \quad 0.16 \quad 0.16 \quad 0.16 \quad 0.04]^T \\ \mathbf{a}_2 &= [0.04 \quad 0.16 \quad 0.16 \quad 0.16 \quad 0.16 \quad 0.30]^T \end{aligned} \tag{7.32}$$

Another issue arises when the pages in the base set form two or more *disconnected subgraphs*. This situation may occur when the query includes terms of an ambiguous (e.g., "Apple", "Java", etc.) or polarized nature, which might result in few, disconnected clusters of pages.

$$\mathbf{E} = \begin{bmatrix} 0 & 0 & 1 & 1 & 0 & 0 \\ 0 & 0 & 1 & 1 & 0 & 0 \\ 0 & 0 & 0 & 0 & 0 & 0 \\ 0 & 0 & 0 & 0 & 0 & 0 \\ 0 & 0 & 0 & 0 & 1 & 0 \\ 0 & 0 & 0 & 0 & 0 & 0 \end{bmatrix} \tag{7.33}$$

Fig. 7.12 Example of
disconnected query result
pages graph

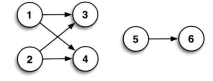

Fig. 7.13 HITS sensitivity to
local topology: (**a**) initial Web
graph; (**b**) addition of node 5

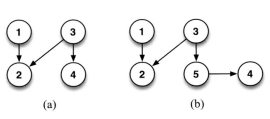

(a)   (b)

As the power method applied in Eqs. (7.25) and (7.26) finds the principal eigenvector of the authority and hub matrices, the largest connected subgraph is favored. Considering for instance the Web graph of Fig. 7.12, and the corresponding adjacency matrix of Eq. (7.33), the calculated hub and authority scores are unique; the final hub and authority vectors $\mathbf{a}^* = [0\ 0\ 0.5\ 0.5\ 0\ 0]^T$ and $\mathbf{h}^* = [0.5\ 0\ 0.5\ 0\ 0\ 0]^T$ clearly show that the subgraph comprising nodes $[1, 2, 3, 4]$ is favored over the subgraph composed by nodes $[5, 6]$.

The authority and hub scores for all the subgraphs can be identified by extending the calculation to the retrieval of the eigenvectors associated with the *top-n largest* eigenvalues (where n is the number of disconnected components). Note that the calculation of high-order eigenvectors also helps to reveal clusters in the graph structure [85].

7.4.4 Issues in HITS Application

Several studies have shown HITS to be *very sensitive to the topology* of the query base set of pages. Considering the two graphs of Fig. 7.13, it is straightforward to calculate how, although differing for only one node and one edge, the resulting authority and hub vectors are dramatically different: the insertion of node 5 in the graph (something that frequently happens due to website redirection or reorganization) completely zeros the scores of node 5.

The Stochastic Approach for Link-Structure Analysis (SALSA) [222] is an evolution of HITS that includes some of the ideas from PageRank in order to make HITS more resilient to changes in the Web graph by performing a random walk on the bipartite hubs and authorities graph, alternating between the hub and authority sides. The random walk starts from some authority node selected uniformly at random. The random walk then proceeds by alternating between backward and forward steps. Several works have thoroughly studied the impact of the selection of the initial base set of pages on the performance of HITS and SALSA [261]: to improve the effectiveness of such ranking algorithms, the neighborhood graph should incorporate only a small subset of the neighbors of each result, and it should avoid edges that do not touch result pages. Additional nodes and edges can be included using consistent sampling techniques.

The graph expansion performed during the first step of the HITS algorithm also makes the method very sensitive to malicious behaviors. A source of problems may come from malicious webmasters who try to artificially boost the hub score of their site by adding several *nepotistic* links connecting pages of the same website, or pages of websites managed by the same organization. Note that the same problem might also involuntarily occur when a website is hosted by multiple servers (e.g., en.wikipedia.org and it.wikipedia.org). To mitigate the effect of nepotistic links, such low-value edges in the graph should not be taken into account (or, at least, deemphasized) when computing hub/authority scores. The original HITS paper of Kleinberg simply discards links connecting pages on the same host. Other works suggest alternative solutions based on link weight normalization through *site-level* weight assignment [46]: if m pages on the same host link to a target page, the corresponding edges are assigned a weight $\frac{1}{m}$.

The initial graph expansion step introduces another class of problems, mainly related to the fact that locality of contents typically works in a very short radius: within a small number of links, the probability that all nodes have the same topic as the starting root set vanishes rapidly, and even at radius one, *topic contamination* may occur, when the expansion step might include pages that are very authoritative but *off topic* because of *generalization* (e.g., query "marathon" finds good hubs/authorities for "sports") or *topic drift* (e.g., the inclusion of highly connected regions of the Web graph about an unrelated topic) issues. A straightforward solution proposed in [46] relies on a content-based discrimination of the expanded pages according to the content of the ones retrieved by the query: by computing the term vectors of the document in the root set, it is easy to compute the centroid of these vectors and then prune pages added in the expansion step that are too dissimilar from the centroid. Alternatively, one can embed a more fine-grained analysis of the link anchors on the original HTML resource, or directly exploit the document mark-up structure to discriminate multi-topic pages.

7.5   On the Value of Link-Based Analysis

Put into context, the advent of link-based analysis completely changed the effectiveness of Web search engines. PageRank and HITS (and their variations) are the two most famous ranking systems for which scientific documentation is available. However, during the last 20 years many researchers have evaluated the actual effectiveness of the two classes of methods, with the goal of understanding which one is better.

The huge success of Google seems to leave no space for discussion, but several considerations can be made. As explained at the beginning of the chapter, PageRank's main difference from HITS is the need for an offline calculation of the page

Fig. 7.14 A simple Web
graph to analyze in
Exercise 7.1

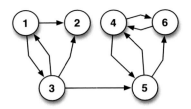

scores, as the method operates on the entire Web graph. However, several works
have shown that a large-scale implementation of HITS is possible ([215] provides
a complete overview on the issue). Another key distinction between the PageRank
and HITS methods is that PageRank lacks the notion of *hub*, a dearth that might be
compensated by the nature of the Web, where good hubs tend to acquire a lot of
inlinks, thus becoming authorities as well [85].

But what about effectiveness? Several studies tried to compare the two classes of
algorithms using the information retrieval evaluation measures described in Chap. 1,
with interesting results. In [337], a comparison between the PageRank value provided by the Google Toolbar and a score calculated according to the indegree of
pages showed similar performance. In [295] it was demonstrated that structure-
independent features, combined with a page's popularity, significantly outperformed
PageRank. Reference [263] offers a large-scale evaluation of the effectiveness of
HITS in comparison with other link-based ranking algorithms, in particular Page-
Rank and indegree, showing that indegree of Web pages outperforms PageRank,
and is about as effective as HITS authority scores (while HITS hub scores are much
less effective ranking features).

7.6 Exercises

7.1 Given the Web graph in Fig. 7.14, compute PageRank, hub, and authority scores
for each of the six pages. Compare and justify the obtained page orders.

7.2 Given the keyword queries "iPhone" and "Samsung Galaxy", compare the result lists of at least two Web search engines. Highlight the results that should be
considered good authorities, and the ones that could represent good hubs, motivat-
ing your choice.

7.3 Recalling Sect. 7.3.5, list and comment a set of at least ten possible ranking
"signals" that could possibly be used by a search engine, and discuss how they
could be used for "black hat" search engine optimization.

7.4 Describe three application scenarios for graph analysis techniques that can be
used to improve the performance of an information retrieval system.

Chapter 8
Recommendation and Diversification for the Web

Abstract Web search effectiveness is hampered by several factors, including the existence of several similar resources, ambiguous queries posed by users, and the varied profiles and objectives of users who submit the same query. This chapter focuses on recommendation and diversification techniques as possible solutions to these issues. Recommendation systems aim at predicting the level of preferences of users towards some items, with the purpose of suggesting ones they may like, among a set of elements they haven't considered yet. Diversification is a technique that aims at removing redundancy in search results, with the purpose of reducing the information overload and providing more satisfying and diverse information items to the user. Both techniques, although typically "hidden behind the curtains" to the end user, result in dramatically enhancing user satisfaction with the performance of retrieval systems.

8.1 Pruning Information

Information overload, generated when a search request produces too much information, is a critical aspect of Web search (as already mentioned in Chap. 4). It can depend on various factors:

- The *user request is ambiguous*, subject to multiple interpretations; this leads to the need of collecting the responses for each of those interpretations, and then deciding which ones to show to the user.
- The *retrieved results are very similar to each other*; thus, if one of them is relevant, all of them are relevant.
- The *profiles, purpose, and activities of users are extremely diverse*, and therefore the algorithms that build responses to search queries must take into account all of them.

One way to address these problems is to apply classification and clustering techniques, as described in Chap. 4. However, classification and clustering are not the only options for helping users to face the information overload; this chapter focuses on recommendation and diversification techniques. Recommendation systems deliver information targeted to the specific user profile, while result diversification removes items from result sets. Both techniques, although typically hidden to the

S. Ceri et al., *Web Information Retrieval*, Data-Centric Systems and Applications, DOI 10.1007/978-3-642-39314-3_8, © Springer-Verlag Berlin Heidelberg 2013

end user, result in dramatically enhancing user satisfaction with the performance of retrieval systems.

8.2 Recommendation Systems

Recommendation systems, also known as recommenders, aim at predicting the level of preferences of users towards some items, with the purpose of suggesting to them the ones they may like, among a set of elements they haven't considered yet. The final objective is to increase the business volume by proposing goods available for sale or by providing targeted content that is likely to be appreciated by the user and thus will generate online traffic or transactions.

Recommenders are widely applied in e-commerce [309], advertising, and Web search for providing users with results that are somehow correlated with their current activity and still would not appear as direct responses to their questions. The popularity of recommenders is due to the impressive impact they have on increasing online sales (and transactions in general). Some studies estimate increases in the order of 5–20 %, which could be substantial in large businesses [42]. However, the impact of recommendations is not always easy to estimate completely, as they can have indirect effects in the medium and long term that are difficult to track (e.g., because customers do not immediately react to the recommendation but then remember it and buy the products later on, or because of repeated purchases over time, due to one initial recommendation) [107].

Recommenders have also shown some weaknesses, e.g., when they produce very focused and homogeneous result sets. Thus, their results need to be mediated by considering aspects such as diversity, novelty, and coverage [217].

8.2.1 User Profiling

Recommenders can be seen as a branch or extension of customer relationship management (CRM) systems: on one hand, they exploit all the available information on the customer, and on the other hand, they provide additional insight into the user behavior. Based on the available information, they provide personalized insight to the customer. In this sense, recommenders are also personalization tools.

The core asset for recommenders is therefore the *user profile*, which includes explicit demographic information regarding the customer (e.g., age, sex, location, family status, and job) and customer behavior tracking (both online and in the physical world, through loyalty programs, expense tracking, and so on). In some cases a full profile is available, while in other cases the system can rely only on partial information. For instance, when the user is not explicitly logged on to the website or search engine, only limited behavioral information is available and no demographic details can be exploited. Obviously, the richer the profile, the more precise recommendations will be.

When calculating their results, recommenders are able to exploit not only the profile of the current user, but also profiles of other (similar or dissimilar) users as well as a group or clusters of users. Furthermore, recommenders rely on the information available on the items to be recommended. The extent of the contribution of each factor depends on the recommendation approach adopted, as described next.

8.2.2 Types of Recommender Systems

Recommendation techniques can be classified in two main categories:

- *Content-based recommendation* exploits information and features of the items, with the aim of recommending to users a set of items that have some similarity or relationship with items they bought or examined in detail during current or past online activities.
- *Collaborative filtering* exploits similarity between users (i.e., commonalities between user profiles) for suggesting items. It is based upon the assumption that people with the same opinion on an item are more likely to have the same opinion on other items too. Therefore, it is possible to predict the taste of a user by collecting preferences from other similar users.

These two broad categories include several different techniques, which we present next. Hybrid solutions that combine both content-based and collaborative filtering are possible, and experiments show that they actually slightly improve the performance of both.

8.2.3 Content-Based Recommendation Techniques

The main content-based techniques for building recommenders include the following:

- *Search-Based Recommendation.* Starting from a search query and the corresponding result set, search-based recommendation selects and highlights some of these results based on general, non-personalized ranking, e.g., on the number of sales, number of clicks, or number of views. This solution is very simple and only requires some basic statistics on the items; however, it can hardly be considered a recommendation at all, as items are just retrieved in a specific order.
- *Classifiers.* As described in Chap. 4, classifiers can be considered recommendation solutions, as they build classes of items, which can then be recommended to users. They can be combined with other recommendation methods.
- *Category-Based Recommendation.* This technique assumes that each item belongs to one or more categories. When a user picks an item, the system selects potential categories of interest based on the past user activities over items (or over categories), and then selects the top items from the selected categories, which are

recommended. This solution requires that the user's activity be tracked and that the recommendation be somehow personalized.

- *Semantics-Based Recommendation.* This is an extension of category-based recommendation, because it replaces categories of items with arbitrarily complex domain models for describing the semantics of items (e.g., ontologies, vocabularies, and any other kind of conceptual model) and matches these descriptions with semantic models of users [335].
- *Information Filtering.* This technique exploits syntactical information on the items and/or semantic knowledge about their types or categories; the available information spans from unstructured, natural language text to structured descriptions, organized in typed attributes. When a user declares interest in an item, the system also recommends those items that are more similar to it, where similarity is calculated in a way that depends on the type of information. This solution does not need user tracking, but it suffers from two main problems: (i) the recommendation only depends on the description of items and therefore will not change over time to reflect the user's changes of interest; (ii) it does not work when the items are not well described.

Content-based recommenders already feature several interesting behaviors. First, they allow recommendation of *new, rare, or unpopular items*, when needed, because the matchings and predictions are based on item descriptions and not on statistics of usage or preferences. Analogously, they are able to recommend items to *users with unique tastes*, again because no large statistics on preferences are needed. Finally, they are extremely interesting because they are able to *provide explanations of why items were recommended* by listing content-features that caused an item to be recommended. Therefore, content-based recommenders do not suffer from the typical problems of collaborative techniques, such as cold start, sparsity, and first-rater problems (described in the next section).

However, they also suffer from some intrinsic disadvantages: they require that content be encoded through meaningful and measurable features; and they need to represent users' tastes as a learnable function of these content features. Their main weakness is that they are unable to exploit quality judgments of other users (unless they are explicitly included in the content features).

8.2.4 Collaborative Filtering Techniques

Collaborative filtering is the most adopted kind of recommendation nowadays on the Web. Using collaborative filtering requires maintaining a database of as many user ratings of items as possible. The methods finds those users whose ratings strongly correlate with the current user, and then recommend items which are highly rated by similar users.

The simpler scenario for collaborative filtering is to consider Boolean values as ratings. This could represent the fact that users explicitly declared their preference

Table 8.1 Purchase matrix: customers and items they purchased in the past

Customers vs. Items:	Item1	Item2	Item3	Item4	Item5	Item6
Customer 1		X	X			
Customer 2	X			X	X	
Customer 3	X				X	X
Customer 4		X	X	X		
Customer 5	X				X	X
Customer 6		X	X			X
Current customer	X				X	

for an item, or that they actually bought an item, or simply that users clicked or visited that item.

A simple case is shown in Table 8.1, where an X represents the fact that a user bought an item. The recommender must understand which items should be recommended to the current customer, represented as the last row of the table. To do so, the system compares the customers' behavior with the current one and decides which purchases are most likely for the current user. In the example, the current customer behaves similarly to customers 2, 3, and 5 (they all bought Item1 and Item5). Therefore, the system will suggest other items they bought. In particular, it will strongly suggest Item4 (purchased by two customers) and also Item6 (purchased by one customer). In general, the more a customer is similar to the current one, the higher is his weight in the suggestion; and the higher the number of similar customers that bought a certain item, the higher will be the item's rank in the suggestion.

This approach is extremely powerful and efficient, because it is able to deliver very relevant recommendations. Obviously, the bigger the database and the recorded past behavior set, the better the recommendations. However, due to the size of the data, it is difficult to implement and it ends up being a resource- and time-consuming solution. Furthermore, the technique suffers from well-known problems:

- *Cold Start.* At the beginning of the system lifecycle, when the machine is turned on, there is no record of any user interaction or rating. This makes it impossible to calculate any recommendation.
- *First Rater.* New items that have never been purchased would never be recommended by the basic technique described above.
- *New User.* Also, new customers who have never bought anything cannot receive any recommendation because they cannot be compared to other customers.
- *Sparsity.* In general, the purchase table (as well as any rating table or click table) is very sparsely populated: the typical situation is that, upon a warehouse of tens of millions of products, each user may have purchased a few items at most, may have visited a few tens, and may have explicitly rated or commented on just one or two. This makes an effective implementation more difficult.

To address these issues, clustering techniques can be applied over the users, thus aggregating users that behave similarly and hence defining and storing the common behaviors through a representative selection of actions. This reduces the size of the user set to be considered in the analysis.

Each cluster will then be assigned the typical preferences, based on preferences of customers who belong to the cluster. Thus, customers within each cluster will receive recommendations computed at the cluster level, which is much more efficient. Since customers may belong to more than one cluster, clusters may overlap; recommendations are then averaged across the clusters, weighted by participation. This makes recommendations at the cluster level slightly less relevant than at the individual level, but on the other hand it improves performance and reduces the size of the dataset.

8.3 Result Diversification

Diversification is a technique that aims at removing redundancy in search results, with the purpose of reducing the information overload and providing more satisfying and diverse information items to the user.

8.3.1 Scope

Diversification is used for various purposes. One option is to use it for addressing the problem of ambiguity of search queries [105]. For instance, a search query that only contains the keyword "Jaguar" is considered as highly ambiguous, as it may have several interpretations: it could refer to the jaguar as an animal, or to the Jaguar car brand, or to the Jaguars football team. While top-k results of the search could refer to only one of these interpretations, diversifying results and showing items for different interpretations could increase the probability of addressing the actual user need and thus increase his satisfaction.

Diversification is also useful in rebalancing the impact of personalization, which aims at tailoring results to the preferences of a specific user [208]. Diversification can complement preferences (which could overly limit the results) and provide more balanced and satisfying results (as in [290]).

8.3.2 Diversification Definition

Diversification aims at maximizing a quality criterion that combines the *relevance* and the *diversity* of the objects seen by the user. The problem of diversifying items can be expressed as selecting a subset S of cardinality k over a given set X of cardinality n, so as to maximize an objective function that takes into account both relevance and diversity of the k elements of S.

Three commonly used objective functions are *MaxSum, MaxMin*, and *Maximal Marginal Relevance (MMR)*, which implicitly maximizes a hybrid objective function whereby relevance scores are summed together, while the minimum distance between pairs of objects is controlled. Result diversification is a well-investigated topic; [111] provides a survey of existing approaches, while [141] discusses a systematic axiomatization of the problem.

8.3.3 Diversity Criteria

Diversity can be defined according to different criteria, such as content, novelty, and coverage:

- *Content-based definitions* consider diversification as the problem of choosing k out of n given points, so that the minimum (or, more frequently, average) distance between any pair of chosen points is maximized (this is known as the p-dispersion problem, studied in operations research).
- *Novelty-based definitions* aim at distinguishing between novel and redundant items when perusing a list of results. In this case, the level of redundancy of each document is measured with respect to its similarity to all previously retrieved items, calculated as a distance.
- *Coverage-based definitions* aim at covering as many different interpretations as possible of the user query. This can be done by defining taxonomies or categories of queries, and maximizing the coverage of that categorization by the retrieved results.

8.3.4 Balancing Relevance and Diversity

As stated above, the objective of diversification is to balance relevance with a diversity factor.

The relevance of an item with respect to the query can be expressed quantitatively by means of a user-defined (possibly non-monotone) *relevance score function* normalized in the $[0, 1]$ range, where 1 indicates the highest relevance. Then the result set is sorted, e.g., in descending order of relevance.

Analogously, for each pair of items, it is possible to define a diversity measure normalized in the $[0, 1]$ interval, where 0 indicates maximum similarity. Based on the notion of diversity, it is possible to address the problem of extracting from the result set of a query the topmost relevant and diverse combinations.

Let $N = |\mathscr{R}|$ denote the number of items in the result set and $\mathscr{R}_k \subseteq \mathscr{R}$ the subset of items that are presented to the user, where $k = |\mathscr{R}_k|$. Let q denote the user query and let τ be an item in the result list.

We are interested in identifying a subset $\mathcal{R}_k$ which is both *relevant* and *diverse*. Fixing the relevance score $S(\cdot, q)$, the dissimilarity function $\delta(\cdot, \cdot)$, and a given integer k, we aim at selecting a set $\mathcal{R}_k \subseteq \mathcal{R}$ of items, which is the solution of the following optimization problem [141]:

$$\mathcal{R}_k^* = \underset{\mathcal{R}_k \subseteq \mathcal{R}, |\mathcal{R}_k| = k}{\operatorname{argmax}} \ F\big(\mathcal{R}_k, S(\cdot, q), \delta(\cdot, \cdot)\big) \qquad (8.1)$$

where $F(\cdot)$ is an objective function that takes into account both relevance and diversity. Two commonly used objective functions $F(\cdot)$ are MaxSum (Eq. (8.2)) and MaxMin (Eq. (8.3)), as defined in [141]:

$$F(\mathcal{R}_k) = (k-1) \sum_{\tau \in \mathcal{R}_k} S(\tau, q) + 2\lambda \sum_{\tau_u, \tau_v \in \mathcal{R}_k} \delta(\tau_u, \tau_v) \qquad (8.2)$$

$$F(\mathcal{R}_k) = \min_{\tau \in \mathcal{R}_k} S(\tau, q) + \lambda \min_{\tau_u, \tau_v \in \mathcal{R}_k} \delta(\tau_u, \tau_v) \qquad (8.3)$$

where $\lambda > 0$ is the parameter specifying the trade-off between relevance and diversity.

Another objective function, closely related to the aforementioned ones, is *Maximal Marginal Relevance* (MMR), initially proposed in [75]. Indeed, MMR implicitly maximizes a hybrid objective function whereby relevance scores are summed together, while the minimum distance between pairs of objects is controlled.

Solving problem (8.1) when the objective function is given in (8.2) or (8.3) is NP-hard, as it can be reduced to the minimum k-center problem [143]. Nevertheless, greedy algorithms exist [141], which give a 2-approximation solution in polynomial time, whose underlying idea is to create an auxiliary function $\delta'(\cdot, \cdot)$ and iteratively construct the solution by incrementally adding combinations in such a way as to locally maximize the given objective function. Note that for $\lambda = 0$ all algorithms, including the ones based on MMR, return a result set which consists of the top-k combinations with the highest score, thus neglecting diversity.

8.3.5 Diversification Approaches

Within the existing corpus of work on diversification, a broad distinction can be made between the contributions that focus on diversifying search results for document collections (e.g., [291]) and those that concentrate instead on structured datasets, i.e., datasets that contain structured objects [229, 317, 342]. Diversification in multiple dimensions is addressed in [109], where the problem is reduced to MMR by collapsing diversity dimensions in one composite similarity function.

One important application scenario is to apply diversification of structured results sets as produced by queries in online shopping applications. The work [342] shows how to solve *exactly* the problem of picking K out of N products so as to minimize an attribute-based notion of similarity and discusses an efficient implementation technique based on tree traversal. Relevance is also considered, by extending the differentiation to scored tuples: the choice of K items minimizes the measure of similarity but at the same time produces only K-subsets with maximal score (in some sense, giving precedence to relevance with respect to (w.r.t.) diversity). The base of the proposal is the availability of a domain-dependent ordering of attributes (e.g., car make, model, year, color, etc.), which is used to build a prefix-tree index supporting the selection of the subset of objects with maximal score end minimal similarity.

Finally, the work [229] investigates the diversification of structured data from a different perspective: the selection of a limited number of features that can maximally highlight the differences among multiple result sets. This work is thus oriented to identifying the best attributes to use for ranking and diversification.

8.3.6 Multi-domain Diversification

Multi-domain diversification is a diversification approach applied to combinations of objects pertaining to different domains instead of single objects. Since it is applied to complex and combined objects, multi-domain diversification typically requires one to apply combined categorical and quantitative diversity functions.

In the case of categorical diversity, multi-domain diversification can be partially reduced to prefix-based diversification, by choosing an arbitrary order for the categorical attributes used to measure combination diversity. The multi-domain search applications addressed here assume for simplicity unambiguous queries and thus a fixed interpretation.

Two different criteria can be applied to express the similarity (or symmetrically, the diversity) of combinations:

- *Categorical Diversity*: Two combinations are compared based on the equality of the values of one or more categorical attributes of the tuples that constitute them. based the same object the two combinations. As a special case, categorical diversity can be based on the key attributes: this means evaluating if an object (or subcombination of objects) is repeated in the two combinations. Intuitively, categorical diversity is the percentage of tuples in two combinations that do not coincide on some attributes. When these attributes are the key, categorical diversity can be interpreted as the percentage of objects that appear only in one of the two combinations.
- *Quantitative Diversity*: The diversity of two combinations is defined as their distance, expressed by some metric function calculated over a combination of attributes coming from any element in the combination.

In both cases, for each pair of combinations, it is possible to define a diversity measure normalized in the [0, 1] interval, where 0 indicates maximum similarity.

The evaluation of diversity in multi-domain result sets must assess the ability of a given algorithm to retrieve useful and diverse tuples within the first K results in the query answer. One possibility is to apply an adapted α-*Discounted Cumulative Gain* (αDCG) to measure the usefulness (gain) of a tuple considering the subtuples and the relations involved in a multi-domain query, based on its position in the result list and its novelty w.r.t. the previous results in the ranking. This approach is proposed for instance in [56].

8.4 Exercises

8.1 Define the list of features that should be used by a content-based recommender on a website that sells books (such as Amazon.com).

8.2 Present the possible applications of recommenders in the context of online advertising.

8.3 Provide an intuitive description of the difference between the objective functions `MaxSum` (Eq. (8.2)) and `MaxMin` (Eq. (8.3)).

Chapter 9
Advertising in Search

Abstract The business model of search engines mainly relies on advertising for economic sustainability. This chapter provides the foundational concepts of online advertising and search advertising by describing the main advertising strategies that can be adopted (brand advertising and direct marketing), providing the basic terminology of the field, and discussing the economic models (pay per impression, pay per click, pay per conversion) and the auction mechanisms that are implemented in advertising.

9.1 Web Monetization

Internet users are acquainted with the possibility of getting a plethora of services and information sources online, without paying a fee or purchasing a costly subscription. The case of search engines is paramount: users access most of their online experience through search engines (or also Web portals) at no cost. This is very convenient, especially for end users who don't want to spend money on their basic Web navigation experience. However, this should not lead to the naive belief that things are working like this by chance or because they represent an acquired right for Web users. Applications, data, and services are able to be up and running on the Web only if they are economically sustainable, which means that they have a business model that lets them acquire enough economic support to sustain their cost of development, maintenance, and hosting.

Depending on the business model adopted by the application, different ways of acquiring economic resources can be enacted, but for search engines, sustainability is only obtained through *site monetization*, the process of converting the user traffic of a website into monetary revenues. This is typically obtained through *online advertising* (or Web advertising), which is now considered as a big business sector, with total spending estimated in the order of 40 to 50 billion dollars in the United States alone.

9.2 Advertising on the Web

Online advertising is a specific implementation of the marketing strategies of companies that consider it relevant to show their brand or products to an audience of potential customers. Marketing strategies typically cover two main objectives:

- *Brand Advertising*: Aims at making a brand widely recognized in the general public. This does not entail immediate impact on sales and does not solicit immediate action from the targeted audience. The purpose is instead to create a general image of trust, reliability, and elegance (or other positive feeling) in customers, that is retained for a long period of time and that will become useful when the person is in need of one of the products covered by that brand.
- *Direct Marketing*: Consists of explicit mention of products, prices, and offerings that moves the audience towards reacting with an immediate purchase. This includes limited-time offers, where customers are pushed to act immediately.

Both objectives can be addressed through Web advertising, although different means are typically used. Indeed, while originally the Web was perceived for being a good medium mainly for direct response marketing, trends show that it is currently used more and more for brand advertising too. Direct marketing currently covers around 60 % of online advertisement expenditures. Two main solutions exist for Web advertising, as exemplified in Fig. 9.1:

- *Display advertising*, represented by commercial messages in the form of images, videos, or text that appear on Web pages or search engines for the purpose of promoting brands, products, or services. They are purchased by advertisers by leasing an area in the Web page, where the message is shown. Examples include online banners and also commercial pictures, videos, and backgrounds in news websites and portals. An example is shown in Fig. 9.1(b). Display advertising can be sold with a guaranteed delivery (GD) agreement, in which case a given amount of impressions are granted to be delivered in a certain timeframe; this is the strategy applied for the large and expensive marketing campaigns of famous brands. As an alternative, advertising can be sold with non-guaranteed delivery (NGD), typically in the case of smaller scale and budget campaigns, in which prices and allocations are assigned through auctions. Displaying can be targeted through contextual awareness (i.e., showing the banners or ads in relevant pages) or through behavioral targeting (i.e., taking into account the behavior and activities of the user so far).
- *Search advertising*, consisting in showing advertisements (ads) within search result pages. These are also known as sponsored search results, as opposed to organic search results. The ads are not shown massively to all the search users, but are displayed based on the search query issued by the user. The challenge of the search engine is to display an appropriate mix of ads and organic search results. An example of a search result page including a large number of ads is shown in Fig. 9.1(a).

(a) search advertising examples

(b) display advertising example

Fig. 9.1 Examples of search advertising (**a**) and display advertising (**b**), highlighted by *thick-bordered boxes*

9.3 Terminology of Online Advertising

This section summarizes the jargon of online advertising, which is fundamental for understanding the field. The main technical definitions are the following:

- *Page View.* The visualization of a complete HTML Web page in response to a request from a client (browser). This definition has become somehow fuzzy with the advent of rich Internet applications and Asynchronous JavaScript and XML (AJAX), which allow pages to be partially refreshed and/or reshaped. However, the basic concept still holds.
- *Impressions.* The number of times an ad is displayed to users, independently of whether it is clicked or not. Each visualization counts for one impression. Each page view might generate several ad impressions (one for each displayed ad).
- *Click-Through.* The number of clicks received by a specific link. In advertising, the links of interest are the ones connected to a displayed ad.
- *Click-Through Rate (CTR).* Defined as the ratio between the number of clicks received by an ad and the number of times it has been displayed. CTR is typically expressed as a percentage:

$$CTR(a) = \frac{\#\ ClickThroughs(a)}{\#\ Impressions(a)} \tag{1}$$

  Typical CTR values have fallen from 2 %–3 % in the 1990s to current 0.2 %– 0.3 %.
- *Landing Page.* The Web page that is visualized to the user when he clicks on an online advertising. The purpose of a landing page is to definitely catch the attention of the user by providing the basic information of interest for him.
- *Target Page.* The Web page where the publisher of the ad wants to direct the user. For instance, the user can be redirected to the target page after performing some clicks on a different page.
- *Conversion.* The business objective of the ad publisher. It consists of convincing the user to perform the desired action, and therefore to convert his interest into actual business value. Typical conversions may consist in buying a product, filling in a form, providing some feedback, or registering to a service.
- *Conversion Rate.* The ratio between the number of conversions and the number of page visits obtained through the ad. The conversion rate is typically expressed as a percentage and is defined as:

$$CR(a) = \frac{\#\ Conversions(a)}{\#\ Visits(a)} \tag{2}$$

- *Tracking Code.* A technical piece of code (usually JavaScript) that is added to the site pages to track the clicks and the navigation steps of users, in order to get good analytics on their behavior. The path of a user, from the ad to the navigation in the landing and target pages down to the final conversion, is often referred to as a *conversion funnel*.

Search advertising activities can be monetized in three main ways:

- *Cost per Click (CPC)*. Consists of paying the ads based on the number of clicks they receive. The advertiser pays a certain amount for every click. This is the de facto standard model for search advertising.
- *Cost per Impression (CPI)*. Consists of paying for ads based on the number of impressions. Given the low amount per impression that is usually negotiated, the actual measure is given as cost per thousand impressions (also known as cost per mille (CPM)). This is the customary payment method for display advertising.
- *Cost per Action (CPA)*. Consists of paying only for ads that are clicked and finally lead to a desired action from the user, such as a commercial transaction or conversion.

In principle, the above payment models could be applied by fixing a price for the ads (for each click, impression, or conversion, respectively) and then letting each advertiser pay according to the models. However, no online advertising provider offers such an option. All the online advertising business is run through auction-based mechanisms, where advertisers bid (according to the payment model selected by the advertising service provider) to get their ads published.

9.4 Auctions

Online advertising, and more specifically search advertising, is basically sold only through auctions. The auction mechanism is based on some assumptions:

- Search engines have a limited number of available slots for ads in each page, and advertisers compete to acquire those slots by bidding on them.
- Advertisements that are related to the search criteria of the user have a much higher probability of being clicked and give the user the impression of a higher value of the overall search system.
- Not all the slots are equally valuable: ads are displayed in ranked lists, and higher positions in the list are typically more valuable because they tend to receive more clicks (i.e., they have a higher CTR).
- Clicks may have different values for different advertisers, depending on the expected conversion rate and conversion values (e.g., vendors selling expensive products are willing to pay more for the ads).

Considering these assumptions, auctions take the following form:

- Advertisers bid for specific search keywords; with the purpose of having their ads displayed in pages containing search results to those keywords;
- Higher bids win better (i.e., top) positions in ad rankings, while lower bids may result in ads shown at the bottom of the rankings, or not shown at all.

One important point, given the small amount of money associated to each bid and the extremely high number of bids, is to grant that the auction mechanism will be efficient. Two issues are crucial for having efficient auctions:

- The first issue is the estimation of the click probability of each ad [106, 135, 136].
  Several models correlate the click probabilities to various parameters, such as the
  position of the ad, the relative position with respect to the other ads currently
  shown, and so on. The most commonly used model is the *cascade* model, which
  assumes that the user scans the links from the top to the bottom in a Markovian
  way [9, 193]. These models introduce *externalities* in the auction, whereby the
  click probability, and therefore an advertiser's profit, depends on which other ads
  are shown in conjunction.
- The second crucial issue is the design of auctions that provide desirable properties, such as incentivizing truthful reporting (also known as incentive compatibility), i.e., the fact that advertisers bid for the real value of the ad, and not a higher
  or lower amount; and ensuring efficiency of the allocation, which maximizes the
  sum of utilities of all participants, also called the social welfare.[1]

Auctions for display advertising are even more intricate, because they consist of
interactions between many classes of actors, including publishers, advertisers, ad
networks, and ad exchanges; e.g., see [10, 138, 140, 214, 260].

Several auction mechanisms have been devised over time, for the purpose of covering practical requirements of specific usage scenarios or with the aim of granting
better properties. The next sections describe the most well-known ones, in the context of search advertising.

9.4.1 First-Price Auctions

The first and simpler auctions that have been devised are *first-price auctions*. The
term first-price comes from the fact that the winner of the auction must pay the
amount he was bidding, i.e., the first-highest (maximum) bid of the auction. First-
price auctions can be implemented in two main ways: as *English auctions*, where
bids are open (i.e., everybody knows the values of all the bids) or publicly called
(and thus are open again); or as sealed-bid auctions, where bids are not known to
other bidders. While the former type is actually used in some real-life cases, it is
not practical (and thus never applied) to online scenarios. For first-price sealed-bid
auctions, all concealed bids are submitted and then compared. Each bidder can bid
only once, and the person with the highest bid wins the good or service and pays the
amount of his bid to the seller. This means that bidders submit valuations based upon
a supposed value of the good and their own willingness to pay; they do not engage
in direct competition with other bidders. A variant to these auctions is the possibility
of setting a reserve price, i.e., a minimum amount that is needed for selling the good.
The seller reserves the option of not selling the good if that amount is not reached
in the bidding phase.

---

[1]These problems and properties are studied in the field of *mechanism design*, a branch of game
theory.

These kinds of auctions give the bidders incentive to bid lower than their valuation of the item and therefore are not incentive compatible. The reason is that, by bidding at their real valuation, statistically bidders will never have a profit. Indeed, if they bid truthfully and win the auction, they pay exactly the true value of the good (with no profit), and if they lose the auction, again they gain nothing. Therefore, to make a profit (at least sometimes), they need to bid a lower price. This leads to bids with no equilibrium, in the sense that people tend to continuously offer more and more, and then to drop offers down to initial price.

Another practical problem in search advertising is that of "free riders," i.e., advertisers that have their ads always displayed and never pay. This game is played by known brands or services that always bid very high (and thus get published), knowing that they are not going to get any click on the ads (and thus will not pay), because users do not perceive the immediate utility of the link, but the brand still gets huge advantages in terms of visibility. For instance, a marketplace like eBay can advertise its offers for every search keyword; users will seldom directly click on the ads, but this strategy will raise awareness of the fact that anything can be found on eBay. For the above reasons, first-price auctions are no longer applied to online advertising scenarios.

9.4.2 Second-Price Auctions

Second-price auctions, also known as Vickrey auctions after Columbia University professor William Vickrey who studied them in 1961, are a type of sealed-bid auction where bidders submit bids without knowing the bids of the other people in the auction, and in which the highest bidder wins, but the paid price is the second-highest bid. This type of auction gives bidders an incentive to bid their true value for the item, because if they win they are granted some revenue (the difference between their bid and the next one). The result is that bidding the true value is always the best option, regardless of what others do. Despite being widely studied, Vickrey auctions are not usually applied in their original formulation, mainly because that formulation was conceived for one single good, which in advertising means one single available slot.

Therefore, for a sponsored search, the most common format is the *generalized second-price (GSP) auction* [116, 341], which takes into account the possibility of having several slots available. In this auction, the allocation of an ad is based on a combination of the bid and a quality score, the latter being a function of the CTR (the exact function is typically kept secret by the advertising provider). The GSP assigns the slots in (decreasing) order of the bids and lets each bidder pay the next (lower) bid. This means that, for every slot, payment is not based on the bid of the advertiser itself, but on the next highest bid to encourage truthful bidding. Basically, the effect is that payment does not really depend on the bid, but on the position where the ad is allocated.

While this auction works well in practice, it has been shown that the mechanism is not incentive compatible [116]; i.e., bidding truthfully is not a dominant strategy. This means that typically there is an incentive for advertisers to misreport their value, as in first-price auctions. Edelman et al. [115] studied the locally envy-free equilibrium of the GSP game, which corresponds to a stable assignment, i.e., a situation in which nobody wants to swap position and payment with anybody else.

GSP has been proven to be reasonably stable, but also not to have a dominant strategy. The best known mechanism that addresses this issue is the *Vickrey–Clarke–Groves* (VCG) auction, which has been extended to the sponsored search setting [264]. The allocation is identical to GSP (i.e., it always computes efficient allocation), but the payment is slightly different in order to guarantee incentive compatibility. In particular, the payment of the advertiser is calculated by the effect of its presence on the expected utility of other advertisers (called the marginal contribution). In practice, the bid for displaying an ad becomes the maximum amount of money the bidder is willing to pay, while the actual paid amount is under the control of the advertising provider.

All major search engines apply GSP for advertising auctions. However, some corrections are needed to address the practical requirements posed by the many usage scenarios. The main corrections are as follows:

- *Revenue-Based Slot Assignment.* This variant is based on assigning a probability of click to the advertiser (based on the CTR of his ads, used as a measure of quality of the ads), and thus estimating the actual probabilistic revenue of the search engine. Allocation of rank positions is done based on this measure, instead of bids. This solves the problem of free riders, because their ads will get a very low CTR and will thus drop down in the ranking. Overall, this is a huge advantage for the search engine (a rough estimate is that it can typically get at least 30 % higher revenues for each page impression).

- *Minimum Price.* Search engines always fix a reserve price for the advertising auctions, in order to also get a reasonable income in the case of not particularly valuable keywords.

- *Slots Availability.* Since search result pages are quite diverse based on the keywords, kind of results obtained, the size and kind of access device, and personalizations of the interface, a different number of slots can be available in different result pages. This must be accounted for in the allocation.

- *Quality of Landing Page.* The landing page represents a relevant aspect for an ad, because it makes the user feel either rewarded or frustrated when clicking on the ad. As we said, search engines aim at giving the perception of utility of the advertisement, and therefore consider the quality of the landing page as an element in the valuation of an ad. In particular, the merit of the landing page is considered in terms of the relevance of its content with respect to the keywords of the advertiser bidding.

- *Positional Constraints.* Search engines typically define statically some constraints on the position of certain kinds of ads, again to improve the perceived quality of the user and to increase the revenue.

9.5   Pragmatic Details of Auction Implementation

Although the theoretical models and their properties have been thoroughly studied, the concrete implementations face a number of additional challenges related to user behavior, lack of data and analytics, and so on. The main issues are summarized here:

- *CTR Estimation.* Although the entire auction mechanism is based on CTR (and/or derived expected revenue), there is an intrinsic cold start problem related to knowing the actual value of the CTR, which corresponds to a simple question: How can the system know the CTR (and revenue) of ads? The basic solution is to infer an estimate from past performance.
- *CTR of New Ads.* CTR estimation may be extremely difficult for new ads or new advertisers. In this case, the choice of showing or not showing the ad could be critical: on one hand, not showing a new ad could result in not discovering high-performing ads; on the other hand, showing the new ad could result in losing the chance of showing good ads with reliable revenue estimates. In that case, the system tries to quickly reach a reliable estimate by applying an "explore and exploit" strategy, whereby new ads are shown sparsely with the aim of exploring their potential, and then estimates are increasingly refined.
- *Externalities.* As already mentioned, the CTR of an ad is not a variable that depends only on intrinsic properties and qualities of the ad. It also heavily depends on several external factors (externalities), such as the combined presence/absence of other ads, position of other ads, quality and position of organic search results, navigation and search history of the user, and so on. Some works have explored the combined potential of (and the attention gained by) sponsored and organic search, but the problem of taking into account externalities in general is far from being solved.
- *Click Fraud.* CTR is a measure that, in its simpler implementation, is very much subject to possible fraud perpetrated by advertisers, end users, and advertising providers. Advertisers may alter the CTR by clicking themselves on relevant ads. By clicking on adversarial ads they make opponents pay on pages you are not interested in; by automatically generating page impressions and never clicking on opponents' ads, they make their CTR go down, and thus are able to bid lower. Users can be fraudulent too, when they show contextual advertising spaces on their own pages and they click themselves on the ads in order to get the money from the advertising providers, or when they automate page impressions to increase the income of display advertising. Overall, fraud represents a relevant share of CTR (rough estimates are in the order of 10–20 %, also boosted by automatic techniques). That is why advertising providers do their best to mine fraudulent clicks, and do not pay for them.
- *Trusting the Advertiser.* To avoid the problem of CTR fraud, auctions can be performed using a pay-per-conversion strategy. However, this implies that advertisers honestly report to the advertising provider the conversions/actions that are obtained through ads. This can be done through tracking code chunks, but again fraud is possible by adding cheating code in pages to misreport the actions.

- *Trusting the Advertising Provider.* The same happens with the pay-per-click strategy on the provider side: advertisers must trust the provider and the number it reports regarding the clicks of the users (although limited cross-checks can be performed by analyzing the source of incoming users to sites).
- *Sparse Data.* Considering the above issues, we can say that available data for calculating reliable estimates of parameters for the models are currently rather sparse. This makes it hard for both advertisers and advertising providers to set up optimal strategies.
- *Market and Monopolies.* The current state of online advertising is very close to a monopoly, in the sense that very few providers control a huge share of the market and they typically control both search and advertising. This is under the careful analysis of antitrust laws and enforcement bodies, and monetary fines and practical constraints are being considered for violators. Examples of constraints include requiring search engines to highlight more clearly the distinction between sponsored results and organic results.
- *Attention Span.* Despite all attempts at capturing the user's attention, advertising has still caught limited user action. The reason is that the human mind is highly adaptable and able to quickly learn how to cut off content not considered useful. This is known as advertising blindness, i.e., the ability of users to selectively and unconsciously remove sponsored content from their attention span (exactly as they do with TV or real-world poster advertising) [43]. This is not solved by increasing the quantity of ads shown. It might actually be beneficial to show less advertising instead, as data show a risk of falling CTR with excessive exposition of ads.
- *Mobile Advertising.* A lot of attention has recently been given to advertising on mobile devices (within free apps, but also on mobile versions of websites and search sites). This is highly valuable because of the peculiar advantages provided by mobile interfaces to marketers. These include: more precise context awareness (including geo-located position, user profile, and so on); higher monetization of users' time, because mobile devices are often used for short-term needs; and more focused user attention on a limited area and with fewer distractions or get-away options in the navigation. These aspects, together with the much more limited available screen space, make mobile ads more valuable (and thus also more expensive).

9.6 Federated Advertising

Nowadays users increasingly demand complex queries that integrate results from a broad spectrum of data sources. Specialized search engines support the publishing and integration of high-quality data for specific domains (extracted from the deep Web or curated data repositories), and thus are able to produce improved results with respect to traditional search engines. In particular, *federated search systems*

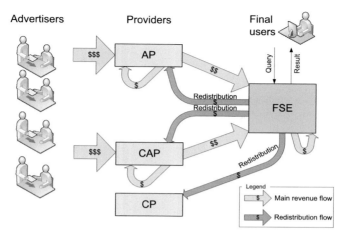

Fig. 9.2 A model of federated advertising

operate on interconnected data sources, offering to their users the impression of a single system. Such systems, reviewed in Chap. 11, respond to the following scenarios:

- *Multi-domain search*, where users submit complex queries about different topics or items at one time. The multi-domain query is split into a number of single-domain queries, each one addressed by a domain-specific content service provider. Results are then produced by combining the various domain-specific items.
- *Meta-search*, where users submit domain-specific queries to a system integrator, which in turn routes the query to many other systems which compete to produce the best answer; results are collected, merged, and reranked before being shown to the user.

   This setting imposes new economic models to deal with the revenue sharing mechanisms; the actors in such scenarios are shown in Fig. 9.2. The federated search system aggregates information from different types of sources, and should provide each of them with adequate compensation. The data providers belong to three categories:

- *Content providers* (CPs) provide datasets with high information value for the users because of their precision, completeness, detail, and/or coverage. Content is the core intellectual property for the CP and is costly to produce.
- *Content and advertising providers* (CAPs) provide content while at the same time serving their own ads (as opposed to using third-party services). Typically, these websites would allow third parties to publish their search results by requiring the federated search engine (FSE) to show (some of) their ads too.
- *Advertising providers* (APs) offer only advertisement services. They publish ads from a set of advertisers, and their goal is to provide targeted ads to users who visit partner websites.

In addition to the three classes of providers, the system involves two classes of stakeholders: *end users*, who visit the FSE website, submit their queries, read the results, and possibly click on the shown advertisements; and *advertisers*, who provide (and pay for) the ads to the APs and CAPs.

A revenue redistribution system is based on the following assumptions/requirements: (i) the only revenue sources of the system are the advertisers, who pay the APs and CAPs for getting their ads published (e.g., with a pay-per-click model); (ii) a different payment or compensation mechanism should be possible for each provider; (iii) in order to grant adequate compensation to every provider that takes part in the FSE's activity, an appropriate redistribution of the revenue may be necessary.

The basic work on redistribution mechanisms was proposed by Cavallo [79], who proposes an (asymptotically) optimal mechanism and thus redistributes as much value as possible when the number of actors goes to infinity (it is not possible to redistribute all the value in the general case [147]) with a computationally efficient algorithm. Another possible redistribution mechanism is that of [151], which allows one to redistribute values in an undominated way according to a given priority over the actors. Only a few works have addressed the problem of revenue sharing mechanisms in search applications [82, 89].

The work [62] extends [82] with an ad hoc revenue sharing mechanism for FSEs, using *mechanism design* [264] to introduce a new payment mechanism that redistributes the yields from advertising to produce adequate monetary incentives to all the actors in the system. The aim is to produce an allocation that maximizes the total revenue of the system. In order to achieve that, the mechanism needs to elicit private information from the providers about the individual ads (e.g., values). However, providers could misreport their true information if they gain more by doing so, thereby hindering the mechanism to find the allocation maximizing the total revenue. APs and CAPs give a portion of the revenue received by the advertisers, as defined by the *payment function* of the mechanism (as in [82]), to the federated search system, which in turn redistributes a portion of the received revenue, as per the *redistribution function* of the mechanism, to the actors. The proposed redistribution function is an extension of the work presented in [79]. In particular, the novelty of our redistribution function is that it is parametric, with a weight assigned to each provider's class, and asymptotically optimal, i.e., it is not possible to redistribute more without creating incentives for some providers to misreport.

9.7 Exercises

9.1 Study some good definitions of conversion in the following cases: e-commerce website, social network, blog for a product company, website of a non-profit entity.

9.2 What would be the most sensible advertising model for mobile applications that have space for displaying only one ad at a time? Which information could be used to better target the ads?

9.3 Design a federated advertising solution for a Web blog where the author wants to integrate banner advertising, contextual ads, and third-party content from a pure content provider and a content and advertising provider.

Part III
Advanced Aspects of Web Search

Chapter 10
Publishing Data on the Web

Abstract During the last 20 years a number of techniques to publish data on the Web have emerged. They range from the inexpensive approach of setting up a Web form to query a database, to the costly one of publishing Linked Data. So far none of these techniques has emerged as the preferable one, but search engine rich snippets are rapidly changing this game. Search engine optimization is becoming the driving business model for data publishing. This chapter surveys, in chronological order, the different techniques data owners have been using to publish data on the Web, provides a mean to comparatively evaluate them, and presents an outlook on the near future of Web data publishing.

10.1 Options for Publishing Data on the Web

As the Web was growing to become the preferred platform for document publication, data-intensive organizations such as national statistical agencies[1] and industries such as e-commerce websites[2] started offering Web forms to search their databases. By 2001, this phenomenon was so vast that it deserved a name of its own: it was called [44] the *deep Web*, because at that time search engines were unable to reach it—in contrast, they were only able to crawl the *surface* of the Web.

For example, consider a company that sells tickets on the phone and maintains a database of upcoming events. When it decides to move its business to the Web by selling tickets online, it can choose between two extreme solutions:

A open a simple website offering a Web form that can be directly used for searching the database;
B create a website where each event has a dedicated page with its own distinguishable IRI,[3] whose content is extracted from the database.

---

[1]For example, http://www.census.gov/ online since 1996.

[2]For example, http://www.amazon.com/ online since 1995.

[3]The internationalized resource identifier (IRI) generalizes the uniform resource identifier (URI) that generalizes the uniform resource locator (URL) used on the Web to identify a Web page. While URIs are limited to a subset of the ASCII character set, IRIs may contain characters from other character sets including Chinese, Japanese, Korean, Cyrillic, and so forth. The IRIs are defined by RFC 3987.

S. Ceri et al., *Web Information Retrieval*, Data-Centric Systems and Applications,   137
DOI 10.1007/978-3-642-39314-3_10, © Springer-Verlag Berlin Heidelberg 2013

Table 10.1 The *five-star* rating system for open data proposed in 2010 by Tim Berners-Lee [45]

Rating	Reason
★	Available on the Web in any format, including an image scan of a table, with an open license, to be open data
★★	Available as machine-readable structured data (e.g., Excel format instead of image scan of a table)
★★★	Available as machine-readable non-proprietary format (e.g., CSV instead of Excel)
★★★★	In addition to three-star rating, available with IRIs that can be used by others to point to the published data items
★★★★★	In addition to four-star rating, the publisher linked its data items to other items already published on the Web to provide context

At the early times of the deep Web, crawlers could navigate the website built as in solution B following the IRIs, while they could not crawl sites built as in solution A, as it was not clear what to write in the Web form. Although the investment for solution A was smaller than the one for solution B, only B could bring to the company customers from search engines. This situation is partially evolving, as search engines are becoming smarter in accessing deep Web content (see Sect. 10.2). This initial example shows that publishing data on the Web has many trade-offs; the more the publisher invests in making published data easy to consume, the more data will be accessed and reused.

A simple and powerful system to discuss the trade-offs of the various methods for publishing data on the Web is to use the *five-star* rating system for open data, proposed in 2010 by Tim Berners-Lee [45] (see Table 10.1), and apply it to all publishing methods (and not just open data).

Data which is published on the Web in any format (including the image scan of a table), with a clear statement that it is published with an open license, is granted one star. A publisher should use one of the existing open data licenses (such as PDDL,[4] ODC-by,[5] or CC0[6]) to indicate that published data is open and can be used. Note that consumers can already benefit from one-star open data: they can look at it, print it, store it locally, enter it into any other system, change it as they wish, and share it with anyone.

One-star data requires either manual work or a dedicated wrapper for data extraction. When instead a software system can read the data format, the data deserves two stars, without putting additional work on the publisher. Data which are published in open formats, e.g., the comma-separated values (CSV) format, deserve the three-star rating, but data owners may point out that open formats are more expensive because

[4]http://opendatacommons.org/licenses/pddl/.

[5]http://opendatacommons.org/licenses/by/.

[6]http://creativecommons.org/publicdomain/zero/1.0/.

(a) converting from their format to an open format has a cost, and (b) maintaining data both in the source format and in open format may cause inconsistency.

Data owners may understand the importance of network effects [339] and trust the approach of *linking and being linked* for increasing the chances of becoming a reference dataset in the open data context (or for reducing the risk of failing, i.e., not being reused in the open data context). In this case, they may pay the extra cost of using IRIs to identify both data and schema items, thus earning a four-star rating; consumers can link to data from any place (on the Web or locally) and they can combine the data safely with other data. Finally, the data deserves the full five stars if the publisher links the data to other data items already published on the Web. In particular, such linking requires the publisher's effort in understanding if another data owner has already published a data item (or a schema item) that refers to the same real-world entity (or to the same concept/relationship).

Going back to the introductory example, both solution A and solution B deserve a one-star rating because they are publishing data in HTML format (which is easier to scrape than an image, but still not a machine-processable data format), but in solution B each event has its own page that can be bookmarked and linked. An event published as in solution B can rapidly jump to a five-star rating by adding to the HTML format machine-processable tags (e.g., those proposed by the microformat hCalendar presented in Sect. 10.4) to indicate the title, the venue, the start date, and the duration of the event.

The rest of the chapter surveys, in chronological order, popular approaches that have been adopted in the last decade to publish data on the Web: the deep Web (Sect. 10.2), Web APIs (Sect. 10.3), microformats (Sect. 10.4), RDFa (Sect. 10.5), and Linked Data (Sect. 10.6). Finally, in Sect. 10.7, the five-star rating system is used to compare these five approaches, to discuss why they became popular among website owners in a given historical moment, and to forecast the near-future development of this field.

10.2 The Deep Web

With the term *deep Web* M.K. Bergman [44] identifies all the information which is hidden behind the query forms of searchable databases. This phenomenon started in the early days of the Web (the mid-1990s) and it is still common, because it gives value to final users at a very limited cost for the data owner.

A conceptual view of the deep Web is well illustrated in Fig. 10.1, published by H. Bin et al. in [158]. The graph of cross-linked documents (the circles) sometimes terminates with a query form that *provides access* to a database. A Web crawler can reach the query form, but then it faces four problems: (a) distinguishing the query forms which give access to databases from other query forms, used for other purposes (e.g., subscribing to a service); (b) inputting content to the form in order to access the database; (c) interpreting the results; and (d) if possible, obtaining permanent links to the data items hidden behind the query form.

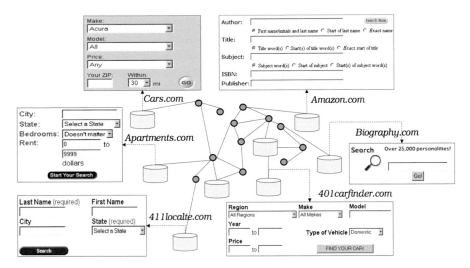

Fig. 10.1 The conceptual view of the deep Web proposed by H. Bin et al. in [158]

Automatic solutions to problems (a) and (c) have been largely investigated; two
seminal works on these two topics are [96], which proposes an approach to automatically discover search interfaces on the Web, and [118], which suggests a technique
to recognize and delimit records in search results. These methods require a large
number of examples in order to learn the patterns.

Practically speaking, these problems remained unsolved until the proposal of
OpenSearch[7] by Amazon in 2005. OpenSearch consists of:

- an Extensible Markup Language (XML) file that *identifies and describes a search
  system* in terms of available query interfaces, example queries, license of the results, and other metadata (see Fig. 10.2);
- a *query syntax* that specifies the signature of a search system (see boldfaced code
  in Fig. 10.2) in terms of search terms, start page (for paginated results), and result
  formats (a choice among HTML, RSS, Atom, and OpenSearch format);
- a *format for sharing results*;
- an *auto-discovery* feature to signal the presence of a search plug-in link to the
  user.[8]

---

[7]http://www.opensearch.org/.

[8]Modern browsers have a search box (e.g., Firefox 18 has it in the top right corner) with which
users can issue search requests to their preferred search engines without opening the search engine
page. The set of search engines reachable from the browser is extensible. When the user is on a
Web page that contains an OpenSearch service description (e.g., amazon.com), the auto-discovery
feature of the protocol allows the browser to detect the search engine interface, and the user can
add it to the set of search engines directly invocable from the browser.

```
<?xml version="1.0" encoding="UTF-8"?>
<OpenSearchDescription xmlns="http://a9.com/-/spec/opensearch/1.1/">
 <ShortName>Event Search by A.com</ShortName>
 <Description>Use A.com to find upcoming events.</Description>
 <Tags>upcoming events</Tags>
 <Contact>admin@A.com</Contact>
 <Url type="application/atom+xml"
  template="http://A.com/?q=searchTerms&amp;pw=startPage?&amp;format=atom"/>
 <Url type="application/rss+xml"
  template="http://A.com/?q=searchTerms&amp;pw=startPage?&amp;format=rss"/>
 <Url type="text/html"
  template="http://A.com/?q=searchTerms&amp;pw=startPage?"/>
 <Query role="example" searchTerms="Lady Gaga" /
 <Developer>A.com Development Team</Developer>
 <Attribution>
    Search data Copyright 2012, A.com, Inc., All Rights Reserved
 </Attribution>
 <SyndicationRight>open</SyndicationRight>
 <AdultContent>false</AdultContent>
 <Language>en-us</Language>
 <OutputEncoding>UTF-8</OutputEncoding>
 <InputEncoding>UTF-8</InputEncoding>
 </OpenSearchDescription>
```

Fig. 10.2 An OpenSearch description of the search interface published using solution A in the example presented in Sect. 10.1

At the time this chapter was written, almost one million websites had adopted OpenSearch[9] and around 4 % of the top 10,000 sites of the Web (e.g., amazon.com, imdb.com, and nature.com) used OpenSearch to describe their query forms, which give access to their databases, and the results returned by those query forms.

For addressing problem (b), a crawler must guess search terms. In the early 2000s, this was difficult; later, the XML element Query in OpenSearch (see Fig. 10.2) was proposed to offer a partial solution to this problem by allowing publishers to list significant search queries. Nowadays, the availability of comprehensive and open knowledge bases such as YAGO (see also Sect. 12.6) has largely simplified this task, as deep Web crawlers (e.g., Google's deep Web crawler [236]) can look up in those knowledge bases. Deep Web crawlers use complex and sometimes site-specific techniques.

Problem (d), the availability of a distinct permanent link to each data item, used to be a severe one in the late 1990s, but with the appearance of Web 2.0 *permalinks*[10] it almost disappeared. In the early days of the Web, many publishers were using the HyperText Transfer Protocol (HTTP) POST method to send search terms to a Web service, which in turn was accessing the database and assembling the result page using HyperText Markup Language (HTML). If the query form uses the HTTP POST method, the returned data items are *not published on the Web*, they are only *accessible through the Web*. When a user fills out the query form and presses the submit button, the form content is sent to a service whose IRI identifies the result page; the

---

[9]Source http://trends.builtwith.com/docinfo/OpenSearch.

[10]A permalink is an IRI pointing to a piece of information published on the text that remains unchanged indefinitely. Most Web 2.0 content syndication software systems support such links.

result pages of two requests issued with different parameters are indistinguishable. In contrast, the HTTP GET method produces result pages which contain the search terms as parameters, and thus the results of two different searches have two different IRIs. A well-designed search interface should use the HTTP GET method and return a list of links that point to the internal pages that describe each result. Google's deep Web crawler [236] is able to crawl information behind both types of search interfaces.

Going back to the example of Sect. 10.1, if the company wants to expose its database of events with a minimum investment (i.e., with a minimal variation of solution A), it should publish a query form that sends the search terms to a service using an HTTP GET, and it should describe the query form using OpenSearch, to allow Web crawlers to identify it and automatically send meaningful queries. The search service should return a result page that lists links to the Web pages where each event is described. As in solution B, each event would then have its own IRI that could be directly returned by a Web search engine like Google.

10.3 Web APIs

As stated before, HTML query forms are a low-cost method for providing access to high-value databases. However, this approach, even with OpenSearch, can hardly be automated. Writing a program that uses an HTML query form to search a database is not difficult, but parsing results which are intended for human consumption is hard. Starting in the late 1990s, much effort was focused on the standardization of Web application programming interfaces (APIs) in order to improve the data exchange methods over the Web. The first proposal, in the late 1990s, was Web Services [17], inspired by the remote procedure call paradigm and centered on XML for exchanged data encoding. Later on, in the late 2000s, the representational state transfer (REST) paradigm emerged, and the JSON[11] format was largely adopted.[12] This new type of Web API is known today as *RESTful Services*.

Over the years, the preferred technical solution changed, but the number of Web APIs available on the Web steadily increased. On November 26, 2012, A. DuVander reported that the ProgrammableWeb directory hit the milestone of 8,000 APIs,[13] just three months after the directory hit 7,000 APIs.[14] ProgrammableWeb is witnessing an exponential growth in the number of Web APIs that has been published on the Web in the last decade. Figure 10.3 proposes three graphs showing the number of services registered in the ProgrammableWeb directory from 2005 to 2012

---

[11] http://www.json.org/.

[12] Interested readers can read a popular blog post entitled "Web Services Org Folds Up and the REST is History" by R. Irani available online at http://blog.programmableweb.com/2010/11/30/web-services-org-folds-up-and-the-rest-is-history/.

[13] http://blog.programmableweb.com/2012/11/26/8000-apis-rise-of-the-enterprise/.

[14] http://blog.programmableweb.com/2012/08/23/7000-apis-twice-as-many-as-this-time-last-year/.

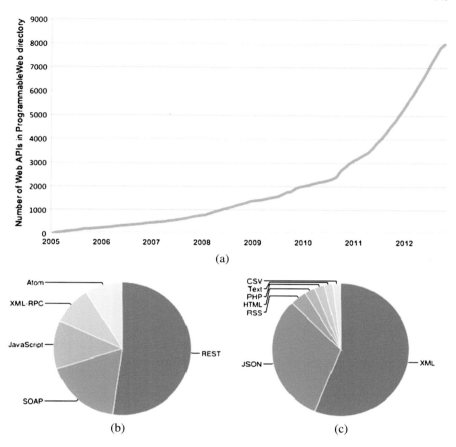

Fig. 10.3   Statistics about Web APIs gathered by ProgrammableWeb

(Fig. 10.3(a)) and the level of adoption of different API protocols (Fig. 10.3(b)) and data formats (Fig. 10.3(c)).

Considering the running example proposed in Sect. 10.1, if the company decides to provide an access to its database using Web APIs, it can provide an event search Web API accessible at the IRI http://api.A.com/events/search whose arguments may be `searchterms`, terms to search in the event title, `location`, a location name or geo-coordinates to use in filtering the search results, and `date`, a date range specified by label (e.g., `Today`, `This Week`, and `Next week`) or an exact range for limiting the list of results. For instance, http://api.A.com/events/search?searchterms=jazz&location=Milano&date=today encodes the request for events whose title contains the word jazz, and that may take place in Milano today. Figure 10.4 shows the JSON syntax of a possible answer, containing three events; the first one is entitled "The Jazz of Marco Bianchi and Paul Banks in Milan", it takes place on January 1, 2013 at 7:00 p.m. in a Milan venue named "Magazzini Gener-

```json
{
"  "events": {
    "event": {
        "title": "The Jazz of Marco Bianchi and Paul Banks in Milan",
        "start_time": "2013-01-02 19:00:00",
        "venue" : {
         "name": "Magazzini Generali",
         "address": "Via Pietrasanta 14",
         "city_name": "Milano",
         "postal_code": "20141",
         "country_name": "Italy"
        },
        "performers": {
        "performer": {
            "name": "Marco Bianchi",
            "Categories": [ "Music", "Jazz"],
            "url": "http://www.linkedin.com/in/marcowhites"
         },
         "performer": {
            "name": "Paul Banks",
            "Categories": [ "Music", "Jazz"],
            "url": "http://myspace.com/paulbanks"
         }
       }
      }
    },
    "event": { ... },
    "event" : { ... }
}
```

Fig. 10.4 An example of a JSON answer that an event search Web API may return

ali", and it has two performers: Marco Bianchi and Paul Banks. The following two events are omitted for space reasons.

In general, data exposed through Web APIs cannot be crawled by Web search engines. Apart from the technical difficulties in crawling data behind Web APIs, many APIs require clients to authenticate themselves and have stringent invocation rate limits that prevent any client from exhaustively crawling the entire database behind the APIs. All together, the information behind those Web APIs forms the modern deep Web. Unlike the deep Web of the 2000s, which was difficult to access because data publishers were not following good Web practices, the deep Web of the 2010s is inaccessible to search engines on purpose. Publishers want to keep control of their data and to allow just a controlled form of usage of their data by Web applications.

Web API publishers do not want search engines to obtain their data, but they welcome *mashups* [365], i.e., the practice of combining data from multiple Web APIs to create a new service. A number of vertical Web search engines are built on Web APIs as mashups. A good example is City Search,[15] an online city guide that

---

[15]http://www.citysearch.com/.

```
<p>On January 1st, 2013 at 7pm, Marco Bianchi and Paul Banks will
perform  jazz  live  in "Magazzini Generali" (Via Pietrasanta 14,
20141, Milano, Italy).<p>
```

Fig. 10.5 An HTML paragraph that describes the event illustrated in Fig. 10.4

provides information about city life throughout the USA. City Search aggregates information provided by Expedia, Hotels.com, Urbanspoon (a restaurant information and recommendation service),[16] Bloglines (an aggregator of top blogs and websites about USA cities),[17] Insider Pages (a search engine for professional services in major USA cities),[18] and MerchantCircle (an online business directory, social business network, and marketing platform).[19] The Web API of the company which was exemplified in this section could fit in City Search to provide information about city events. Readers who are interested in the deployment of services like City Search may refer to [103], which presents issues, solutions, and open problems in realizing this kind of service.

10.4 Microformats

In the mid-2000s, when Web APIs became the preferred way for trading data using HTTP, *microformats* [195] emerged as the approach to openly publish data together with HTML pages. Adopting microformats, a data owner can publish data at the cost of marking up existing pages with machine-processable tags. Microformats offer a way to reuse existing HTML/XHTML tags to convey metadata within information intended for end users.

Consider, for instance, the HTML paragraph in Fig. 10.5 that could have been published by the company adopting solution B in the running example (see Sect. 10.1). The tag p tells a browser to display the sentence as an HTML paragraph. This is enough to properly render the content to an end user, but it does not help when trying to automatically extract information about the event.

Figure 10.6 shows how the paragraph can be annotated using the microformat hCalendar.[20] Each tag span delimits a portion of text as a recognizable data item, and each attribute class describes the type of the enclosed data item. In the example, a span tag, whose class is vevent, is opened at line 2 and closed at line 12. It delimits the entire content of the paragraph as a data item of type *event*. A crawler that understands hCalendar can automatically detect the presence of the event and can start looking for the starting time, description, and location of the event, by using the classes dtstart, summary, and location, respectively. The content of

---

[16]http://www.urbanspoon.com/.

[17]http://www.bloglines.com/.

[18]http://www.insiderpages.com/.

[19]http://www.merchantcircle.com/.

[20]http://microformats.org/wiki/hCalendar.

```
1.  <p>
2.  <span class="vevent"> On
3.  <span class="dtstart" title="2013-01-01T19:00:00+01:00">
4.     January 1st, 2013 at 7pm
5.  </span>,
6.  <span class="summary">
7.     Marco Bianchi and Paul Banks will perform jazz live
8.  </span>in
9.  <span class="location">
10.    "Magazzini Generali" (Via Pietrasanta 14,20141,Milano,Italy)
11. <span>.
12. </span>
13. <p>
```

Fig. 10.6 The HTML paragraph illustrated in Fig. 10.5 with semantic annotations which use the hCalendar microformat

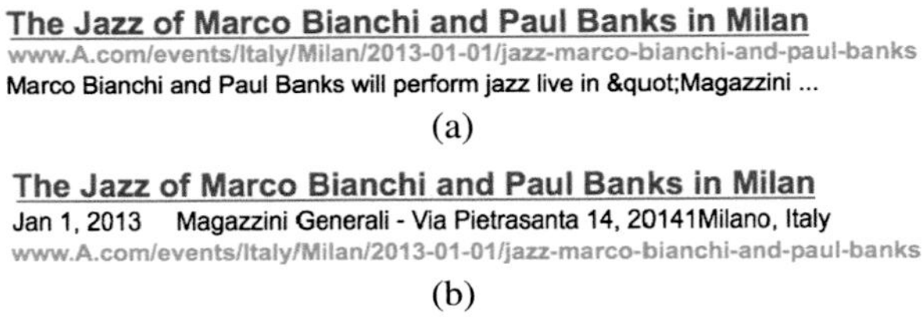

The Jazz of Marco Bianchi and Paul Banks in Milan
www.A.com/events/Italy/Milan/2013-01-01/jazz-marco-bianchi-and-paul-banks
Marco Bianchi and Paul Banks will perform jazz live in "Magazzini ...

(a)

The Jazz of Marco Bianchi and Paul Banks in Milan
Jan 1, 2013   Magazzini Generali - Via Pietrasanta 14, 20141Milano, Italy
www.A.com/events/Italy/Milan/2013-01-01/jazz-marco-bianchi-and-paul-banks

(b)

Fig. 10.7 Comparison of two Google snippets. (**a**) Points to a page without microformats, (**b**) points to a page with microformats as illustrated in Fig. 10.6

the span tag is the value of the event attribute. When the value is not in a machine-readable format, the attribute title can be used to override the content of the span tag. For instance, the value of the attribute title at line 3—the ISO date[21] 2013-01-01T19:00:00+01:00—overrides the content at line 4 of the span tag—January 1st, 2013 at 7pm.

If publishers annotate their pages with microformats (e.g., hCalendar for events), major search engines can better format search results. For instance, Fig. 10.7(a) shows how Google would format a result (more precisely, a *snippet*) that points to a page which is not marked up with hCalendar, while Fig. 10.7(b) shows how Google would format a snippet that points to a page annotated with hCalendar. In the latter case, the search engine can extract the starting date, the summary, and the location, which normally is linked to Google Maps, and can create what is called a *rich snippet*.

Several microformats have been developed through a community process[22] to publish different types of data: hCalendar for events, hReview for individual reviews and ratings, hReview-aggregate for aggregate reviews and ratings, hRecipe

---

[21] The International Standard ISO 8601 specifies numeric representations of date and time. Interested readers can refer to http://www.cl.cam.ac.uk/~mgk25/iso-time.html for a summary.

[22] http://www.microformats.org/.

Fig. 10.8 Example of *rich snippets* that Google uses to format search results that point to pages annotated with microformats

for cooking and baking recipes, hMedia for media information about images, video, and audio, and hCard for people and organization contacts. Figure 10.8 shows more examples of rich snippets offered by Google for events marked up with hCalendar, product reviews marked up with hReview, recipes marked up with hRecipe, and music tracks marked up with hMedia. User studies have shown that rich snippets help users to decide how to select search results; thus, by marking up their pages with microformats the data owners increase their visibility on the search engine. Moreover, rich snippets can provide direct links to the content that a data publisher wants to sell, thus decreasing the number of clicks between content discovery and the potential financial transaction.

Some microformats are also meant to type links [16]. For instance, RelTag [9] proposes to use the `rel` attribute of the hyperlink tag `a` to indicate that some entry relates to a tag from a given (user-specified) vocabulary. Thus, it is possible to state that the event in Fig. 10.6 is about jazz in a machine-processable way by linking the word jazz to the respective Wikipedia category, as follows:

```
<a href="http://en.wikipedia.org/wiki/Category:Jazz"
   rel="tag">jazz</a>
```

The addition of a RelTag for easing content search and correlation was quite successful, resulting in 20 million pages tagged with RelTag in the first six months after its release [196] in January 2005. Other microformats that belong to this category are Rel-License to assign a license to a Web page, Rel-Nofollow to link in untrusted third-party content, and Votelinks to express support for the linked page.

10.5 RDFa

Microformats have been very successful, due to the attention to usability for the publishers, but in some cases they turned out to be too simple:

- Using more than one microformat per Web page can lead to *name collision*.[23] As reported on the microformats community site,[24] using the attribute `class`, microformat markup can overlap with Cascading Style Sheets (CSS) rules. For instance, a publisher may use in a CSS style sheet the class `url` to change the presentation of a link without knowing that `url` has a special meaning in hCard microformat.
- Microformats are impossible to specialize or generalize. For instance, hCalendar includes only summary, start time, and location. If, for a given type of events (e.g., music events), also the performer should be mandatory, a new microformat must be created from scratch, because *no notion of inheritance* is built in.
- No *microformat parser* exists. Each microformat requires its own parser, because microformats do not distinguish the syntactic layer from the semantic one.

Resource Description Framework in Attributes (RDFa) [5] addresses these three issues and provides a specification for attributes to express structured data in any markup language, in particular XHTML [250] and HTML [6]. While microformats specify both a syntax for embedding structured data into HTML documents and a vocabulary of specific terms for each microformat, RDFa specifies only a syntax and relies on independent specification of terms by others (e.g., schema.org, the broad vocabulary specified by Google, Yahoo!, and Bing in a joint initiative).[25] RDFa allows terms from multiple independently developed vocabularies to be freely intermixed and can be parsed without the knowledge of the specific vocabulary being used.

Figure 10.9 shows how the HTML paragraph illustrated in Fig. 10.5 can be semantically annotated using the RDFa syntax and schema.org vocabulary. As one can see by comparing this figure with Fig. 10.6, the syntax of RDFa uses the `span` tag as do microformats, but it does not reuse the attribute `class` It introduces the new attributes `vocab`, `about`, `typeof`, `property`, `content`, and `datatype`.

The `vocab` attribute (line 2) introduces the namespace of schema.org; all the terms used from line 2 to line 16 belong to this namespace. The `about` attribute (line 3) specifies the identifier of the data item using an IRI. The `typeof` attribute (line 4) states that the data item is of type `Event` (see also Fig. 10.10). The `property` attribute (lines 5, 10, and 13) specifies the properties of the data item using the terms of the schema.org vocabulary. At line 5, it describes the start date of the

---

[23]The term *name collision* refers to the problem that occurs in computer programs when the same name denotes different things; typically it arises when two separate program components are merged. Problems of name collision are commonly addressed by introducing *namespaces* and *prefixes*.

[24]http://microformats.org/wiki/issues#opened_2010.

[25]http://schema.org/Event.

```
 1.<p>
 2.  <span vocab="http://schema.org/"
 3.         about="http://www.A.com/events/Italy/Milan/2013-01-01/
                              jazz-marco-bianchi-and-paul-banks"
 4.         typeof="Event" >
 5.  <span property="startDate"
 6.         content="2013-01-01T19:00:00+01:00"
 7.         datatype="xsd:dateTime">
 8.    On January 1st, 2013 at 7pm
 9.  </span>,
10.  <span property="name">
11.    Marco Bianchi and Paul Banks will perform jazz live
12.  </span>in
13.  <span property="location" typeof="PostalAddress">
14.    "Magazzini Generali" (Via Pietrasanta 14,20141,Milano,Italy)
15.  <span>.
16.  </span>
17.<p>
```

Fig. 10.9 The HTML paragraph illustrated in Fig. 10.5 with semantic annotations using the RDFa (*boldfaced*) and schema.org vocabulary (*italicized*)

event, at line 10 the name, and at line 13 the location. As for microformats, the value of a property is the content of the span tag that can be overridden by the value of the content attribute (line 6). Unlike the example in Fig. 10.6, the data type of the content is not implicit; it can be explicitly stated using the datatype attribute. At line 7, the start date given at line 6 is told to be a valid XML Schema date-time. The typeof attribute can also be used to specify the type of a value of a property as on line 13, where the content of line 14 is stated to be a postal address.

RDFa has been largely adopted in the last years. Two well-known examples of successful adoption are Best Buy[26] and Facebook,[27] whose cases are next reviewed.

In 2010, Best Buy deployed RDFa throughout its website, with the goal of increasing the visibility of Best Buy stores in the search engines that offer rich snippets (see Sect. 10.4). At that time, the company's marketing division noted that finding basic store information, like store locations and opening hours, was difficult; consequently, the company management decided to associate each store with a blog, whose main page was annotated with the store name, address, geo data, and opening hours by using the GoodRelations vocabulary [164]. Figure 10.11 shows the front page of the Best Buy store in Carbondale; invisible metadata are published in RDFa together with information intended for the end user. With such RDFa markings, search engines started to identify each data component and to create rich snippets. Figure 10.12 shows what Google returns if asked for "Best Buy Carbondale". Note that all the metadata entered in the store blog (e.g., the post address and the geo data visible on the right of Fig. 10.11) appear in the box that describes the store

---

[26]http://www.readwriteweb.com/archives/how_best_buy_is_using_the_semantic_web.php.

[27]http://www.readwriteweb.com/archives/facebook_the_semantic_web.php.

Thing > Event
An event happening at a certain time at a certain location.

Property	Expected Type	Description
Properties from Thing		
additionalType	URL	An additional type for the item, typically used for adding more specific types from external vocabularies in microdata syntax. This is a relationship between something and a class that the thing is in. In RDFa syntax, it is better to use the native RDFa syntax – the 'typeof' attribute – for multiple types. Schema.org tools may have only weaker understanding of extra types, in particular those defined externally.
description	Text	A short description of the item.
image	URL	URL of an image of the item.
name	Text	The name of the item.
url	URL	URL of the item.
Properties from Event		
attendee	Organization or Person	A person or organization attending the event.
attendees	Organization or Person	A person attending the event (legacy spelling; see singular form, attendee).
duration	Duration	The duration of the item (movie, audio recording, event, etc.) in ISO 8601 date format.
endDate	Date	The end date and time of the event (in ISO 8601 date format).
location	Place or PostalAddress	The location of the event or organization.
offers	Offer	An offer to sell this item—for example, an offer to sell a product, the DVD of a movie, or tickets to an event.
performer	Organization or Person	A performer at the event—for example, a presenter, musician, musical group or actor.
performers	Organization or Person	The main performer or performers of the event—for example, a presenter, musician, or actor (legacy spelling; see singular form, performer).
startDate	Date	The start date and time of the event (in ISO 8601 date format).
subEvent	Event	An Event that is part of this event. For example, a conference event includes many presentations, each are a subEvent of the conference.
subEvents	Event	Events that are a part of this event. For example, a conference event includes many presentations, each are subEvents of the conference (legacy spelling; see singular form, subEvent).
superEvent	Event	An event that this event is a part of. For example, a collection of individual music performances might each have a music festival as their superEvent.

Fig. 10.10 The schema proposed by schema.org for semantically describing an event happening at a certain time at a certain location

on Google (right side of Fig. 10.12). The result was a 30 % increase in the search traffic to the store blog pages, which accounts for the increasing number of users that locate Best Buy stores using Google.

Fig. 10.11   RDFa annotation in the Web page of one of the Best Buy stores

Regarding the relationship of Facebook and RDFa, in April 2010 Facebook announced Open Graph[28] as a protocol to enable publishers to integrate their Web pages into the Facebook social graph. The protocol includes features such as *Like buttons* and publisher plug-ins. It also includes a simple vocabulary to be used together with a simplified version of RDFa that enables publishers to describe the objects in a page. The Open Graph vocabulary contains the following terms:

- og:title—The title of the object the page is about, e.g., "Marco Bianchi and Paul Banks will perform jazz live" in the example presented in Figs. 10.6 and 10.9;

---

[28]http://developers.facebook.com/docs/opengraph.

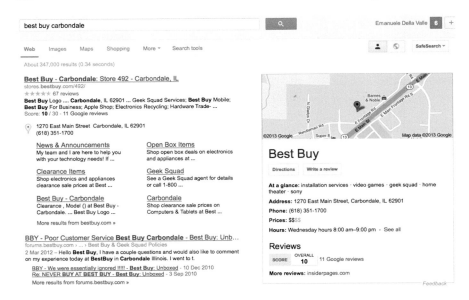

Fig. 10.12   How Google renders the search results for Best Buy stores in Carbondale. Notably, the *box on the right* reports the RDFa annotation shown in Fig. 10.11

- og:type—The type of the object, e.g., "Event";
- og:image—An image URL (absent in the example);
- og:url—The permanent ID of the object, e.g., http://www.A.com/events/Italy/Milan/2013-01-01/jazz-marco-bianchi-and-paul-banks;
- og:description—A short description of the object, e.g., "On January 1st, 2013 at 7pm, Marco Bianchi and Paul Banks will perform jazz live in Magazzini Generali (Via Pietrasanta 14, 20141, Milano, Italy)";
- og:site_name—If the object is part of a larger website, the name which should be displayed for the overall site, e.g., A.com.

Web pages annotated with Open Graph vocabulary can be referenced and connected across social network user profiles, blog posts, search results, and so on. Figure 10.13 shows the growing success of the Open Graph protocol. As of November 2012, 3.3 million sites were using it,[29] showcasing the usefulness of RDFa.

10.6   Linked Data

We note that this chapter presents two families of techniques. Section 10.2 on the deep Web and Sect. 10.3 on Web APIs illustrate techniques centered on giving access to an already existing database, the choice proposed in solution A. Section 10.4 on microformats and Sect. 10.5 on RDFa discuss approaches focused on using the

---

[29]Source http://trends.builtwith.com/docinfo/Open-Graph-Protocol.

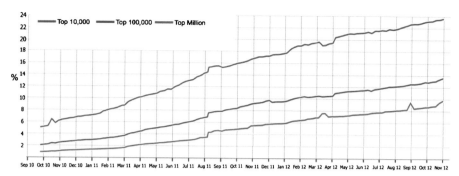

Fig. 10.13 Three million sites are using the Open Graph protocol, which uses RDFa

data in a back-end database to enrich Web pages oriented to end users, the choice proposed in solution B. The Linked Data initiative intends to bridge this gap and provide a comprehensive technology stack that covers both use cases.

The term Linked Data [47, 48] indicates a set of best practices for publishing and linking structured data on the Web. Adopting Linked Data best practices means publishing data on the Web in machine-readable formats, explicitly defining its meaning, linking it to other external datasets, and, in turn, being linked to from external datasets. Linked Data relies on IRIs to identify data items, on HTTP as a simple yet universal mechanism to retrieve resources across the Internet, on RDF [216] as a data model, and on RDFS [63] or OWL [251] as languages for creating vocabularies.

C. Bizer et al. [48] note that Linked Data, being a direct extension of the Web, shares some of its properties:

- It offers a generic platform that can contain any type of data.
- Anyone can publish data.
- Data publishers can choose to represent data in any vocabulary.
- Data items are connected by links, creating a global data graph that spans data sources and enables the discovery of new data sources.

It is beyond the scope of this chapter to introduce RDF, RDFS, and OWL. However, it is worth noting that the RDF data model is a directed labeled graph (see also Sect. 12.2). Representing structured data in RDF is straightforward. For instance, the data of the jazz event illustrated in Fig. 10.9 can be represented in RDF as in Fig. 10.14. The node in the middle of the graph contains the IRI of the event. It is the *subject* of four RDF *statements*. One declares that the IRI represents an event; graphically it is presented linking the central node to the node whose IRI is http://schema.org/Event, labeling the link as type. In RDF terminology the node pointed to by the link is called the *object* and the link the *property*. A second statement declares the starting time using http://schema.org/startDate as property and the value "2013-01-01T19:00:00+01:00" as object. The third and fourth statements assert the name and the location, respectively.

As a result, data is strictly separated from formatting and presentational aspects. Whether the data is encoded using RDFa together with end-user information or is

Fig. 10.14 A representation in the RDF data model (i.e., a *directed labeled graph*) of the jazz event illustrated in Fig. 10.9

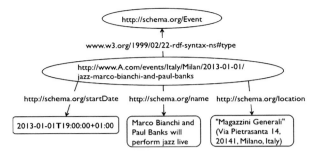

exposed as data through a Web API, it represents the same directed labeled graph. Moreover, data is self-describing. If an application consuming Linked Data encounters data described with an unknown vocabulary, the application can dereference the IRIs that identify vocabulary terms and discover their definitions. In addition, Linked Data is open-ended; as on the Web, data sources can be discovered by following links.

Linked Data best practices have been adopted by an increasing number of data providers over the last three years, leading to the creation of a global data space containing billions of assertions. Figure 10.15 illustrates what is called the *Linked Data cloud*. Each node is a dataset. The presence of an edge between two datasets means that some of the data items in the two datasets are linked. The thicker the edge, the more links exist between the data items in the two datasets.

As Richard MacManus wrote in a post on ReadWriteWeb,[30] "the tipping point for the long-awaited Semantic Web may be when you can query a set of data about someone not too famous (e.g., Modigliani), and get a long list of structured results in return". Amedeo Modigliani is a celebrated painter of the early twentieth century, but he is not as famous as Da Vinci or Picasso. Richard MacManus has challenged the Semantic Web to provide structured answers to the query: "tell me the locations of all the original paintings of Modigliani" (called "The Modigliani Test"): a tool retrieving such locations would be a suitable query language for the Web.

Thanks to the data published in the Linked Data cloud, "The Modigliani Test" was passed by Atanas Kiryakov, in April 2010.[31] He issued a query on FactForge,[32] a product of Kiryakov's company, Ontotext. FactForge was created by collecting, cleaning up, and indexing part of the Linked Data cloud into a single repository in a way that allows efficient and reliable queries. The results were partial, because only some of the museums that house Modigliani's paintings published their data in linked format, but the most famous paintings (i.e., *Reclining Nude, Portrait of Lunia Czechowska with a Fan, Anna Zborowska, Madam Pompadour,* and *Gypsy Woman with Baby*) are present in FactForge.

---

[30]http://readwrite.com/2010/04/15/the_modigliani_test_semantic_web_tipping_point.

[31]http://readwrite.com/2010/04/25/the_modigliani_test_for_linked_data.

[32]http://factforge.net/.

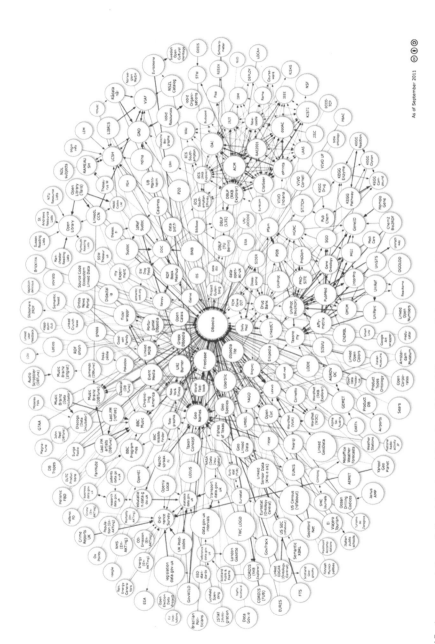

Fig. 10.15 A representation of the datasets published as Linked Data as of September 2011 provided by Richard Cyganiak and Anja Jentzsch on http://lod-cloud.net/

Table 10.2 The five approaches illustrated in this chapter, compared by using the *five-star* rating system illustrated in Sect. 10.1

Approach	Rating	Why
Deep Web	One to three stars	Data items are published in HTML tables; sometimes they can be exported in Excel or CSV formats
Web APIs	Three to four stars	Data items are published in non-proprietary format (e.g., XML or JSON) but are rarely identified by IRI. No outgoing links are generally available
Microformats	Three to five stars	Data items are published in non-proprietary formats using open vocabularies. Often a single data item is published per page, thus implicitly assigning an IRI to the published data item. The frequent usage of link-based microformats earns them a five-star rating
RDFa and Linked Data	Five stars	Publishers are required to identify data items with IRIs, the description of each term in the used vocabulary must be dereferenceable (thus each data item description is at least linked to the vocabulary it uses), and the data model favors the creation of links to other data items

10.7   Conclusion and Outlook

This chapter presents five approaches for publishing data on the Web: the deep Web, Web APIs, microformats, RDFa, and Linked Data. They have all appeared in the last 15 years. Their popularity has fluctuated, but none of them has disappeared so far. They offer trade-offs about who pays the costs between publishers and consumers. If the publisher pays the costs, the data are easier to access for the consumer, but what is the business model of data publishing? The rest of this section uses the five-star rating system illustrated in Sect. 10.1 to compare these approaches (see also Table 10.2) and tries to forecast the near-future development of this field.

The *deep Web*, which provides access to a database using a query form, deserves a one-star rate. Data owners are publishing data with a minimum investment, but the data items are not directly processable. However, if data is released under an open license, then it can be extracted using a wrapper-specific query form and can be republished in formats which can be more easily processed. For instance, the Federal Deposit Insurance Corporation[33] offers a query form[34] to search for failed banks. The result is an HTML table, but the data is open and was made available as a public Google spreadsheet (which deserves a three-star rate) by the Europe's failed banks data party in October 2012.[35] The deep Web will probably keep growing in the next year, but one hopes that all the open data published in this way will be republished

---

[33] http://www.fdic.gov.

[34] http://www2.fdic.gov/hsob/SelectRpt.asp?EntryTyp=30.

[35] http://blog.okfn.org/2012/10/19/data-party-tracking-europes-failed-banks/.

by private or social initiative. The original data owners should carefully monitor this republishing process, as other publishers could make profits by publishing their original data with a better rated approach.

Web APIs deserve a three-star rate, because they use a non-proprietary format such as XML and JSON; they generally miss IRIs to identify data items. At first glance this can appear strange, since mashups need identifiers to join information from multiple Web APIs, but a closer look shows that successful mashups use either exact matching of unambiguous properties of real-world entities (such as postal addresses) or approximate matching of names of popular real-world entities (such as names of famous people or famous sport teams). In the second case, the unavoidable ambiguity carried by names can be resolved by using context. For instance, "Milan", when it appears together with other Italian cities or in a context where data is about location in Italy, can be interpreted as the northern Italy city; when it appears together with other soccer teams, it most likely refers to the sports team; and when it appears together with famous novel writers, it can be matched to "Milan Kundera".

If a Web API also provides identifiers for data items, it can be granted a four-star rate. The earliest examples of this class of Web APIs are the http://www.amazon.com/ ones. In 2008, Amazon.com Inc. introduced the Amazon Standard Identification Number (ASIN) as a 10-character alphanumeric unique identifier for its products worldwide. If a book has an ISBN code, its ASIN code is its ISBN. For all other products sold on Amazon, the ASIN code is generated and guaranteed to be unique worldwide. In recent years, ASIN codes have been picked up by many other websites that link to Amazon products. Major search engines noticed this trend and in the last five years have been adding to their products facilities for publishing data and for querying them through Web APIs. For this purpose, in 2008, Yahoo! rolled out Open Data Tables,[36] a facility to manage data through Web APIs, and Yahoo! Query Language,[37] an expressive SQL-like language that allows for querying, filtering, and joining data across Web APIs. As an answer, in 2009, Google launched Fusion Table,[38] a set of Web APIs for data management with an emphasis on data visualization APIs, and in 2010 Google bought Freebase,[39] an entity graph of people, places, and things, built by the open data community. What will happen next is hard to predict, but most likely the number of Web APIs that provide access to data is going to grow.

Microformats deserve three to five stars. For instance, HTML pages describing events that are marked up with hCalendar, per se, only merit a three-star rating. When each event is published in a single Web page and hence has its own IRI, it earns a four-star rating. If the RelTag microformat is used to link the event to tag taxonomy, microformats can even reach a five-star rating.

A five-star rating is deserved as well by data which is published using *RDFa* and *Linked Data*. Will these W3C standards become more adopted than microformats?

---

[36]http://www.datatables.org/.

[37]http://developer.yahoo.com/yql/.

[38]http://www.google.com/fusiontables.

[39]http://www.freebase.com/.

Fig. 10.16 Google trends reveal that interest in *microformats* after a high peak in 2007 is shading; *RDFa*, *Linked Data*, and *microformats* are capturing a comparable interest; and interest in *rich snippets* is rapidly growing

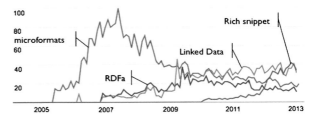

At the time that this chapter was written, Linked Data was supported by publicly funded projects,[40] but RDFa gained a 33 % share of the market of the data published within HTML pages [259].

In this situation it is hard to forecast even the near future. However, a look at Google trends[41] reveals interesting insights (see Fig. 10.16). The interest around microformats after a high peak in 2007 is shading. Linked Data have been gaining popularity since 2007, surpassing (in terms of number of Google's searches) microformats in 2009. In 2012, Linked Data, RDFa, and microformats appeared to have captured a comparable interest. The most interesting trend concerns rich snippets. One hopes that publishers will stop the long-lasting (and sterile) discussion contrasting alternative data publishing approaches and focus their attention on determining which data publishing solution is the most cost-effective in terms of building rich snippets.

10.8 Exercises

10.1 What is the deep Web? What is the origin of this term? Will the deep Web disappear in the future?

10.2 Using hCalendar, annotate the following HTML code:

```
<p>On June 12th, 2013, Cyndi Lauper will stop in San Diego for
"She's So Unusual" (2241 Shelter Island Dr, San Diego, CA,
United States).<p>
```

10.3 Using RDFa and schema.org vocabulary, annotate the following HTML code:

```
<p>On June 12th, 2013, Cyndi Lauper will stop in San Diego for
"She's So Unusual" (2241 Shelter Island Dr, San Diego, CA,
United States).<p>
```

---

[40]The European Commission within the Seventh Information and Communication Technologies Work Programme funded the projects LOD2 (http://lod2.eu) and LACT (http://latc-project.eu) to further push the development of Linked Data technologies, and LDBC (http://www.ldbc.eu/) to establish industry cooperation between vendors of Linked Data technologies in developing, endorsing, and publishing reliable and insightful benchmark results.

[41]http://www.google.com/trends/.

10.4 Using the graphical syntax shown in Fig. 10.9, draw the RDF graph corresponding to the solution of the previous exercise.

10.5 What is the five-star rating system for open data publication? How does it work?

10.6 Why can hCalendar potentially score less than RDFa using the five-star rating system?

Chapter 11
Meta-search and Multi-domain Search

Abstract While search engine technology based on crawling and indexing, presented in Part II, dominates the market, a niche is open for search systems based on data integration technology. These systems either rely on other search engines as sources of information or directly access specialized data sources that are focused on given domains. Interest in such systems is growing with the increase of Web applications which offer simple query interfaces to domain-specific data sources. This chapter overviews the theory of rank-driven data integration and top-k query processing, and then focuses on meta-search and multi-domain search.

11.1 Introduction and Motivation

The focus of this chapter is on higher order search, i.e., the ability to perform several independent search activities over a variety of data sources in response to a query, and then assemble a global result which integrates and ranks the results of the various search activities. In principle, the idea is very powerful, as it consists of using search systems of various natures as building blocks; in practice, obtaining effective results from the concurrent use of different sources or systems is not easy, due to the inherent difficulty of data integration. Two relevant approaches to search system integration are meta-search and multi-domain search.

- *Meta-search* consists of addressing the same query to systems which are knowledgeable *about the same domain*, essentially in order to compare their results, rank them, and present the best ones to users. This idea was very popular, especially during the years 2005–2007, but it has been contrasted for a variety of pragmatic reasons (see, in particular, the report [35], with a recommendation of *directly searching individual search engines to get the most precise results, and using meta-searchers if you want to explore more broadly*). Meta-search has regained momentum in specific vertical sectors, most notably price comparison, where tools such as Trivago (specialized in hotels) and Kayak (specialized in flights) compare the prices of various vendors.
- *Multi-domain search* consists of addressing different queries to systems which are knowledgeable *about different domains*, and then assembling *combinations* of query results, aiming at a global optimum typically expressed as a weighted

sum of the relevance of the objects retrieved by individual data sources, e.g., finding the best combination of hotels and restaurants in a given geographic region and possibly located not too far from each other. Multi-domain search is still a research topic; the relevance of the field is increasing with the growth of available data sources (including *Linked Data* which were discussed in the previous chapter) and with the slow but constant progresses of technologies for data extraction and integration from different sources.

This chapter will briefly overview meta-search and then concentrate on multi-domain search. However, first, we will review the foundational theory of *top-k query processing*, focused on how to efficiently extract the best results from *structured* data sources which are potentially very large.

11.2 Top-*k* Query Processing over Data Sources

A *data source* contains information about real-world objects, such as movies, restaurants, or hotels. Objects in data sources have structured attributes, typically:

- An *object identifier* OID, which must have different values for different objects. When a source can be accessed by fetching objects by their OID, it is said to support *random access*.
- A *score attribute S*, which contains a quality measure associated with every object; scores of different objects can be used for ranking objects and can be aggregated to compute a global score for the composite responses of the result set. A source that can be accessed by fetching objects in descending order of their scoring attribute is said to support *sequential access*.
- Several *value attributes* A_i, denoting domain-dependent properties. When a source can be accessed via predicates on value attributes, it is said to support *attribute-based access*.

Data sources often provide *access limitations*, as they can only be accessed through specific procedures. For instance, a data source with a score attribute might not support sequential access.

Several problems and corresponding methods are defined for extracting data from data sources in an optimal way, based on the different assumptions about their access limitations and on the kinds of query; queries are formulated using classical query languages or their variants which take into account access limitations. Queries typically include a *global score function* which associates a global score to the query results. Users are normally interested only in the query results with the *best* score; therefore, *top-k queries* are computed, where *k* is the size of the result set. Global scores are monotonic functions of the (local) scores associated with each object; conventionally, we consider score maximization as the query objective.

Optimality typically refers to the minimization of the number of accesses required for obtaining the query results, as well as of the time for producing the first *k* results—it is assumed that the methods halt as soon as *k* top results are available

Fig. 11.1 Examples of two
data sources for the
OID-based problem

Source 1		Source 2	
OID	SCORE	OID	SCORE
7	70	1	70
1	60	3	60
3	50	5	50
4	40	4	40
5	30	7	30
2	20	6	20
6	10	2	10

and return them to the enclosing environment. Objects and their scores can also be individually returned as the methods progressively establish that they belong to the result. Two typical problems are considered:

- *OID-Based Problem*: In the presence of several sources for the same objects, find the object ordering that maximizes a global score function. This problem takes advantage of the object identifiers for detecting the same object in the various sources, performing an OID-based join.
- *Attribute-Based Problem*: In the presence of several objects, each extracted from a given source, find the object combination that minimizes a global score function, where a combination is obtained by n-ples of objects which satisfy join predicates on their value attributes.

11.2.1 OID-Based Problem

This problem is characterized by n sources S_i of the same type of object, each with an OID and a score s_i, and a global scoring function $F(s_1, s_2, \ldots, s_n)$. Figure 11.1 shows two sources, each containing objects with OIDs ranging from 1 to 7; the global score function is $F = s_1 + s_2$.

TA Method

The threshold algorithm (TA) method, proposed by Fagin et al. [123], uses both sequential and random accesses, and it assumes that each object is present in every source.

The method performs a parallel sequential access to all the sources, followed by random accesses to those objects in a source S_i which were already retrieved from some other source S_j but not from source S_i. In this way, after the i-th round of parallel access and subsequent sequential accesses, the method accesses h objects from all sources, with $h \geq i$, and computes their global score. An upper bound T (the threshold) represents the global score of unseen objects; this bound is computed by applying the global score function to the objects retrieved at round i with the sequential access. Therefore, the value of T decreases at each round. Objects whose

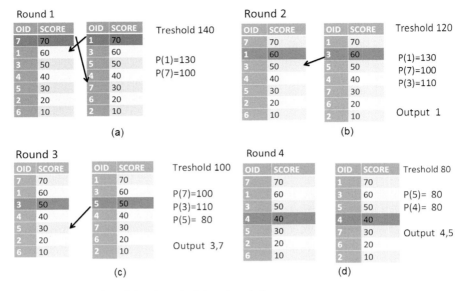

Fig. 11.2 Application of the TA method, iterations 1–4

global score function exceeds T can be returned as query results, ranked by their global score.

An example of application of the TA method is shown in Fig. 11.2(a)–(d). Figure 11.2(a) shows round 1: objects 7 and 1 are sequentially retrieved from sources S_1 and S_2, and subsequently the method requires a random access to object 7 from S_2 and 1 from S_1. T is set to 140, the sum of the top scores (70) for both sources. Thus, after round 1, two objects 1 and 7 are seen ($h = 2$), but none of them has a value exceeding T. Figure 11.2(b) shows round 2: objects 1 and 3 are sequentially retrieved from sources S_1 and S_2, and then the method makes a random access to object 3 from S_1, while no random access is needed for object 1, which has already been extracted from both sources. T is set to 120, the sum of the current top scores (60) for both sources. Thus, after round 2, three objects 1, 3, and 7 have been retrieved, and the score of object 1 is higher than the threshold; so object 1 can be returned in output. Figure 11.2(c), (d) show rounds 3 and 4; after round 4, the output consists of the list of objects $\langle 1, 3, 7, 4, 5 \rangle$, at the cost of 8 sequential accesses and 4 random accesses.

NRA Method

The no random access (NRA) method, also proposed by Fagin et al. [123], assumes the availability of sequential access, but not of random access. Each object is present in every source, and each local score has a smallest possible value (e.g., nonnegative numbers, with 0 as the bottom value).

Round 1   Round 2

OID	SCORE		OID	SCORE		OID	RANGE		OID	SCORE		OID	SCORE		OID	RANGE
7	70		1	70		7	70:140		7	70		1	70		7	70:130
1	60		3	60		1	70:140		1	60		3	60		1	130:130
3	50		5	50					3	50		5	50		3	60:120
4	40		4	40					4	40		4	40			
5	30		7	30					5	30		7	30		OUTPUT 1	
2	20		6	20					2	20		6	20			
6	10		2	10					6	10		2	10			

                      (a)   (b)

Round 3   Round 5

OID	SCORE		OID	SCORE		OID	RANGE		OID	SCORE		OID	SCORE		OID	RANGE
7	70		1	70		7	70:120		7	70		1	70		7	100:100
1	60		3	60		3	110:110		1	60		3	60		4	80: 80
3	50		5	50		5	50:100		3	50		5	50		5	80: 80
4	40		4	40					4	40		4	40		OUTPUT 7,4,5	
5	30		7	30		OUTPUT 3			5	30		7	30			
2	20		6	20					2	20		6	20			
6	10		2	10					6	10		2	10			

                      (c)   (d)

Fig. 11.3   Application of the NRA method, iterations 1–3 and 5

The method performs parallel sequential accesses and then assigns a lower and an upper bound to all the retrieved objects. Both bounds are computed by applying the global score function; if an object has not been retrieved yet from a given source, the lower bound is the smallest possible value at that source, and the upper bound is the last retrieved score for that source. Objects are reported when they are among the top-k results and their lower bound is not below the upper bounds of all other objects, including unseen objects. Note that this method allows reporting objects when they satisfy the above condition, even if their global score is not known.

An example of an application of the NRA method is shown in Fig. 11.3(a)–(e); assume that all scores are nonnegative numbers and therefore the smallest possible score is 0. Figure 11.3(a) shows round 1: objects 7 and 1 are sequentially retrieved from sources S_1 and S_2, and subsequently the method computes their upper and lower bounds, using 0 and 70 respectively as the smallest and the last retrieved score. Figure 11.3(b) shows round 2: objects 1 and 3 are sequentially retrieved from sources S_1 and S_2, object 1 has been retrieved from both sources and so its global score evaluates to 130; this is therefore also its lowest and upper bound. The bounds for objects 7 and 3 must be computed; note that the upper bound for object 7 decreases, as the last retrieved score from source 1 is 60. Given that the lower bound of object 1 is not below the upper bounds of all other objects (including unseen objects whose bounds are (0, 120), it can be reported. Figure 11.3(c), (d) show rounds 3 and 5; after round 5, the output consists of the list of objects $\langle 1, 3, 7, 4, 5 \rangle$.

Fig. 11.4 Comparison of the
TA and NRA methods

OID	SCORE	TA Round	With direct accesses	NRA Round
1	130	2	3	2
3	110	3	1	4
7	100	3	0	5
4	80	4	0	5
5	80	4	0	5
2	30	7	2	7
6	30	7	0	7

Fig. 11.5 Comparison of
NRA (on the *left*) and greedy
(on the *right*) methods

OID	SCORE	Round		OID	SCORE	Round
1	130	2		1	130	2
3	110	4		3	110	3
7	100	5		4	80	4
4	80	5		7	100	5
5	80	5		5	80	5
2	30	7		2	30	7
6	30	7		6	30	7

Trade-Offs

As shown by the two examples (see their summary in Fig. 11.4), TA uses a richer set
of access primitives and thus converges with fewer sequential accesses than NRA;
however, it requires random accesses, which are in general more costly. Note, however, that convergence of NRA could be very slow with certain distributions of score
or top values at specific sources, e.g., with range value distributions that make the
lower bound not so tight.

Considering that the global score function is monotone and that the sequential
access produces objects in descending score order, a pragmatic alternative to top-k
methods is using a *greedy method* which immediately outputs their join results, at
the risk of outputting out-of-order results; the method can be adapted by dynamically reordering results in the user interface. Figure 11.5 shows that the obtained
final greedy ranking approximates the top-7 ranking (with one inversion) and that
results are returned faster (in two cases).

11.2.2 Attribute-Based Problem

This problem is characterized by n sources S_i each with an OID, a score s_i, and
other attribute(s) A_i. Queries over these sources include a join condition, expressed
as the conjunction of join predicates $A_i \theta A_j$, normally equi-join predicates. As an
effect of computing the join condition over the objects extracted from each source,
the query computes *combinations* $\langle o_1, o_2, \ldots, o_n \rangle$ of objects; each combination is
associated with a global score. Queries extract top-k combinations.

Figure 11.6 shows two sources, each containing objects with OIDs ranging from
1 to 7; the join condition is $A1 = A2$ and the global score function is $F = s_1 + s_2$.

Fig. 11.6 Examples of two data sources for the attribute-based problem

Source 1				Source 2		
OID	**SCORE**	**A1**		**OID**	**SCORE**	**A2**
1a	70	3		1b	70	1
2a	60	2		2b	60	2
3a	50	1		3b	50	3
4a	40	4		4b	40	4
5a	30	2		5b	30	2
6a	20	1		6b	20	3
7a	10	3		7b	10	1

Fig. 11.7 Application of the rank-join method

OID	SCORE	A1		OID	SCORE	A2		OID	OID	GLOBAL SCORE	ROUND
1a	70	3		1b	70	1		1a	3b	120	3
2a	60	2		2b	60	2		2a	2b	120	3
3a	50	1		3b	50	3		3a	1b	120	3
4a	40	4		4b	40	4		1a	6b	90	6
5a	30	2		5b	30	2		2a	5b	90	6
6a	20	1		6b	20	3		5a	2b	90	6
7a	10	3		7b	10	1		6a	1b	90	6
								4a	4b	80	7

Rank-Join Method

The rank-join method, proposed by Ilyas et al. [180], is very similar to the NRA method. The main difference is that, as join results are a subset of the Cartesian product of join attributes which in turn depends on the query, no assumption can be made about their existence, and therefore bounds cannot be used. The method performs parallel sequential accesses and computes the join tuples and their global score. The threshold T represents the scores of combinations which have not been formed yet. At each round of sequential access, first the global score function is applied to n combinations, yielding n values T_i; each such combination is formed by taking for $n-1$ sources the highest score value from that source (seen at the first access) and for the one remaining source the last score value seen at that source; then, T is set to the maximum of T_i. Combinations are reported when they are among the top-k and their global score is not below T.

An example of an application of the rank-join method is shown in Fig. 11.7; first, we discuss how the threshold T is computed at each round. At round 1, T is simply 140. At round 2, we have $T_1 = 70 + 60 = 130$, $T_2 = 60 + 70 = 130$, hence also $T = \max(T_1, T_2) = 130$. With similar arguments, T decreases by ten units at each round, until round 7, when it is $T = 80$. Consider now the combinations that are formed when the join condition holds, reported in Fig. 11.7. We describe combinations as pairs of OIDs, and associate them with their global score function. The figure reports as well the round at which combinations can be reported. Note also that other combinations can be formed; however, they cannot be reported due to the existence of the threshold.

HRJN Method

The hash rank-join (HRJN) method, also proposed by Ilyas et al. [180], presents an efficient way of computing the NRA method, based on the existence of hash join structures on the join attributes that are used for fast, direct access to matching tuples within the sources; hence, it is also very similar to the TA method. Thresholds are maintained for each source and also globally. The algorithm includes a decision step for selecting the most convenient source for the next access, based on the current formed combinations and threshold values; this decision step is called the *pulling strategy*, and it can itself be optimized [3].

Search Computing Method

The search computing method [84] computes the top-k combinations by using both sequential and attribute-based accesses. It assumes that each sequential access returns *chunks* of tuples so as to limit the number of accesses to sources (and to optimize the transmission of results), but we will not further discuss this aspect. The method consists of two phases; in the first phase, enough sequential accesses to sources are performed so as to construct a set C of at least k combinations. These are not guaranteed to be the top-k combinations; however, it is guaranteed that the top-k combinations can be formed by the second phase of the method, consisting of several attribute-based accesses. The second phase starts by enumerating *all* the values v_{ij} for the join attributes A_i of source S_i that appear in some combinations of C, then for each such values it considers the attribute B_j of source S_j which is equi-joined to A_i and performs an attribute-based access to S_j on B_j with input value v_{ij}. After all the attribute-based accesses, new combinations are built, and they are ranked in global score order; the top-k combinations are returned. The method includes provisions for managing the failure of the first phase, if there are not enough combinations to answer the query, or in the case when sequential access to one source fails, as its tuples are exhausted.

The search computing method, once applied to the example represented in Fig. 11.7, computes 4 combinations in phase 1, reported in Fig. 11.8(a); therefore, after phase 1 there is a guarantee that the top-4 combinations can be found. Then, the attribute-based accesses in phase 2 (with both attributes $A1$ and $A2$ set equal to values 1–4) allow building additional 8 combinations, yielding a total of 12 combinations, shown in Fig. 11.8(b) in global score order; the first 4 combinations yield the top-4 list. Note that one of the combinations found in phase 1 ends up not being part of the top-4 list.

11.3 Meta-search

Meta-search was originally motivated by the assumption that the Web is too big to be successfully indexed by a single search engine, and therefore the most comprehensive search results can be obtained by routing the same query to several engines.

Fig. 11.8 Phases 1 and 2 of the search computing method

OID	OID	GLOBAL SCORE
1	3	120
2	2	120
3	1	120
4	4	80

OID	OID	GLOBAL SCORE
1	3	120
2	2	120
3	1	120
1	6	90
2	5	90
5	2	90
6	1	90
4	4	80
7	3	60
3	7	60
6	7	30
7	6	30

Today, this motivation is clearly not valid, given that the major search engines cover the entire Web. The current motivation, discussed below, is the finding that major search engines differ in their rankings: their top choices are sufficiently different to motivate their parallel search and top-choice fusion.

In general, a meta-search engine does not contain its own indexes on Web resources. Rather, it contains methods for accessing several search engines in parallel, routing them the query, obtaining the top query results, and integrating them, by removing duplicates and by ordering them according to a global ranking; some meta-search systems instead present their results clustered by the producing search engine. The process of querying multiple search engines is called *federated search*, and it typically entails an initial query transformation and a final result merge and integration; in both processes, linguistic knowledge is essential to generate the appropriate query terms and to recognize duplicate entries.

Another key aspect of meta-search is the ability of offering *faceted search*, i.e., exposing different facets of the same information; each facet could map to given media (e.g., videos or images) and consequently to specialized search engines (e.g., YouTube and Flickr). The query interpretation phase should be focused upon understanding the various facets underlying a given query and then mapping each facet to given search engines; results are then assembled and presented in order of affinity with the given facets.

Kosmix [292] was a very popular meta-search engine during the years 2008–2010; the company was then bought by Walmart and the technology was transferred. Kosmix's approach for exploring data sources is based upon a very large, hierarchical topic taxonomy, reflecting *is-a relationships*, consisting of several millions of entities organized as a directed acyclic graph; beyond is-a relationships, the taxonomy supports other semantic relationships, e.g., *member-of, capital-of*, and so on. The taxonomy was built over a period of three years using both automatic methods and manual curation, and has a few million nodes. Some of the nodes are also associated with the data sources that are most knowledgeable about each node's topic; the system knows thousands of data sources, from very generic (e.g., Google) to very specific (e.g., Epicurus, a website about food and wine). When a query is asked, a *query categorization service* is able to determine the nodes in the taxonomy most closely connected to the query and then select about 10–20 sources which are most suited to answer the query. For instance, a query on "pinot noir" is understood as a kind of wine, and hence a kind of beverage, and also as a kind of grape, related to viticulture, and such nodes are mapped to websites such as Epicurus, FoodNetwork, DailyPlate, and FatSecret (such a mapping is described in [292] and is dated 2009). Results were presented in a two-dimensional layout and classified according to their type (text, audio, video, news, and so on) and rankings—in this way, Kosmix was presenting their various facets. Moreover, a window presented links to other topics related to the query, as extracted from the taxonomy, so as to enable exploratory queries.

Dogpile (www.dogpile.com) is a popular meta-search system connected to a variety of search engines. It gained its maximum popularity during 2007–2009; at the time of this writing, the website efficiently processes queries. Dogpile's website motivates meta-search using a study jointly developed with the Queensland University of Technology and Pennsylvania State University.[1] The study has evaluated about 20 K queries into the four most popular search engines of the time (Google, Yahoo!, Windows Live, and Ask) by comparing the first page of results. It found that only 0.6 % of results are shared by the four systems, while results which appear in just one system account for 88.3 %. On average, 69.6 % of Google first page results were unique to Google, and the percentages for Yahoo!, Windows Live, and Ask were even higher (79.4 %, 80.1 %, and 75.0 %). These data motivate the conclusion that best quality results require accessing many search systems, and not just one. The report also indicates that Dogpile misses in its first page those results that are returned by any two pairs of search engines only in 21.5 % of the queries, and reports the results which are returned by all the search engines 97.9 % of the time.

Other popular systems perform domain-specific meta-search. The most famous system, *Trivago* (www.trivago.com), compares the cost of hotels which are offered by the various hotel booking integrators. Trivago scans over a hundred hotel booking sites and extracts not only the hotel features (descriptions, images, reviews) but also prices, thereby associating each hotel with a number of offers, each with a cost. For instance, hotel Melia in Milano is quoted for 91 euro by Expedia, 99 euro by Bookings and Last Minute, and 107 euro by Olotels (Aug. 2012). The policy of Trivago is to be an intermediary: bookings occur on the pointed hotel booking sites, while Trivago profits from advertising, and charges the *click-through* (whenever users follow a link from Trivago to a hotel booking site, the site pays Trivago). *Kayak* (www.kayak.com) offers a similar meta-search service for flight offers.

Finally, some meta-search systems give to their users the task of selecting the specific categories related to their query: thus, they offer to users a variety of choices of the query domains, but then behave as domain-specific meta-searchers once the category is selected. Among them, *Copernic Agent Personal* (www.copernic.com)

---

[1]The white paper can be retrieved from http://www.dogpile.com/info.dogpl/support/metasearch.

offers a choice of about 20 favorites (including *cars, television, wine*, etc.) and then a few other main categories (*entertainment, news, recreation*) further subclassified, for a total of about 50 domains. Once a category is selected (e.g., top news), the system indicates how many search systems are available and then routes the query to them. Users can follow the query process and find out specifically which sources are addressed and how fast they respond, and see the news with associated ranking being progressively displayed and reranked.

11.4 Multi-domain Search

Several domain-specific data sources about, e.g., entertainment, restoration, accommodation, transportation, small businesses, housing, job offers, and schools exhibit a ranking capability relative to simple queries. Multi-domain search is concerned with queries over multiple domains, e.g., asking for the best combination of hotels and restaurants within the same district of a city. We assume that hotels and restaurants can be accessed using distinct sources, and also that each source can independently rank them, e.g., by decreasing price. The difficulty of the query lies in finding a combination of hotels and restaurants within the same district and possibly satisfying other conditions (e.g., a hotel with at least three stars and an Italian restaurant) such that their total cost is minimized; in addition, the distance between them could be either constrained or also be subject to minimization. Relative to metasearch, multi-domain search adds the complexity of query decomposition, query routing, result assembling, and efficient extraction of ranked results. It uses the top-k query processing theory described in Sect. 11.2, with OID-based methods when data objects are uniquely identified across the various sources (e.g., some Linked Data applications), and attribute-based methods otherwise. Most of the discussion of this section is related to the Search Computing (SeCo) project [83].

11.4.1 Service Registration

Domain-specific data sources are open to public availability through Web services. When services extract ranked data in response to queries, they can be regarded as *search services*; the underlying data ordering technique can be either a simple order by query, e.g., in SQL on top of tabular data, or an arbitrary keyword-based information retrieval method with given precision and recall. In many cases, the ranking is not exposed as a value, as it is simply expressed by the position of each result item in the result set (the ranking is *opaque*).

The peculiarity of Web services is to extract data with given patterns of input–output interaction (i.e., they provide outputs of given format in response to parametric queries whose input also has a given format). Thus, a service call can be issued only when all its input parameters are set equal to a value; we say that its input parameters are *bound*, using a terminology borrowed from logic programming.

In addition, these services normally produce results in chunks of given size; thus, for a given input, several calls may be needed in order to collect the complete result. In general, only a few calls to search services are sufficient to answer a top-k query.

Querying services through their native interfaces is quite laborious; therefore, a multidimensional search application must offer a conceptual structure of the available information, mapping the user's information need over the data produced by queried search services. Several methods and approaches use the annotation of resources against semantic knowledge models (e.g., ontologies) to improve search; see the next chapter and specifically [125] and [139].

The SeCo framework relies on the service registration and annotation method described in [282]. Its main component is the domain diagram (*DD*), an entity relationships diagram which presents a simple semantic view, for simplifying the exploration of information and the definition of search queries by non-expert users. The *DD* represents service properties after the concepts of an external, "off-the-shelf" knowledge base (*KB*); it uses YAGO 2 [166], which integrates GeoNames, WordNet, and Wikipedia. The registration process, described in [285], uses natural language processing (NLP) methods to map field names to attribute names, giving preference to names which are already in the *DD* or to the terms' preferred meaning in the reference knowledge base. Terminological coherence is guaranteed by the fact that each concept in *DD* corresponds to exactly one concept in the *KB*.

Services are represented at three levels of abstraction:

- At the lowest, *physical* level, a *service interface* which exposes input–output fields (parameters) is directly imported according to a wrapping technology that depends on the specific service implementation; fields are typed and may be either atomic or nested, i.e. have subfields of arbitrary depth.
- At the intermediate, *logical* level, every service field is mapped to an entity attribute, yielding to what are called *access patterns*, and one of the entities is chosen as the *focus entity*—typically the most representative for describing the data items returned by the service; the mapping includes a labeling of attributes as either input, output, or *ranking*, to denote the ordering condition applied by the class of services.
- Finally, at the *conceptual* level, all the access patterns which refer to the same focus entity are clustered, and each cluster is called a *service mart*.

An example of a *DD* is shown in Fig. 11.9; for ease of reading, we show just entities and relationships. The *DD* is obtained as a result of the registration of services about movies, restaurants, hotels, fashion shops, and metro stations.

As examples of logical-level services, consider *Movie* with users' scores, *Theater* with daily schedules of movies, and *Restaurant* with ratings; *Theater* and *Restaurant* can be accessed in decreasing distance order from a given initial location—they are simplified versions of classical *Local* services provided by Google or Yahoo!. The three services support sequential access in ranked order.

Fig. 11.9 Example of data dictionary

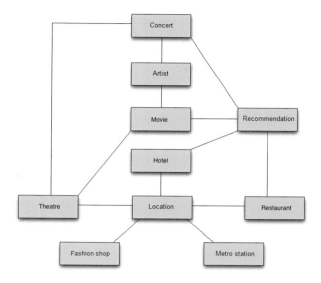

Movie(*Genres.genre^I*, *Country^I*, *Title^O*, *Director^O*, *UsersScore^R*, *Year^O*, *Language^O*, *Actors.name^O*)

TheaterAP(*Date^I*, *BaseAddress^I*, *TName^O*, *TAddress^O*, *Distance^R*, *Showings.title^O*, *Showings.time^O*)

RestaurantAP(*BaseAddress^I*, *Categories.name^I*, *RName^O*, *RAddress^O*, *MapUrl^O*, *Distance^O*, *Rating^R*)

The superscripts *I*, *O*, and *R* denote which fields in the access patterns are marked as input, output, and ranking fields, respectively. Services Theater and Restaurant return instances of theaters and restaurants in ranked distance from the Address which is provided in the input. Fields can be either single-valued or multi-valued; e.g., movies have many actors and may be tagged with many genres, theaters schedule many showings, and restaurants may be tagged with many categories.

11.4.2 Processing Multi-domain Queries

Multi-domain queries span over two or more entities; they can be computed when the corresponding access patterns can be joined. As an example of a query over the sources introduced above, we consider a search for a good and classic American action movie showing in a theater not too far from the user's home address and with a good Mexican restaurant nearby. Below, the query of the running example is expressed in an SQL-compatible syntax.

```
SELECT Movie AS M, TheaterAP AS T, RestaurantAP AS R
WHERE  M.Title = T.Showings.title AND T.BaseAddress = $INPUT_1
     AND M.Genres.Genre = 'Action' AND M.Country = 'USA'
     AND M.Year < 1990 AND  T.Date = today()
     AND R.BaseAddress = T.TAddress
     AND R.Categories.name = 'Mexican'
ORDER BY 0.25 M.UserScore + 0.25 T.Distance + 0.25 R.Rating
     + 0.25 R.Distance
LIMIT 50
```

Variables prefixed as $INPUT have a value provided by users at query execution time; in the example, the value assigned to the variable $INPUT_1 is bound to the field T.BaseAddress; the ORDER BY clause gives equal weight to the four ranking factors; the LIMIT clause restricts the query result to the first 50 combinations. A multi-domain application, e.g., for leisure planning on a mobile device which implicitly provides the T.BaseAddress, could be associated with the above query, typically with several variables to be bound at execution time reflecting user's choices (such as the genre of the movie, the category of the restaurant, different weights of ranking factors, and possibly max global distances and costs).

Invocability of an access pattern in a query expresses the fact that all its input fields are bound to actual values, either directly by means of input parameters or constant values, or indirectly by using the content returned by calls to other access patterns. Formally, the access pattern AP_i is (recursively) said to be *invocable* when, for every input field $I_h \in AP_i$, the query Q contains either a selection predicate $I_h = const$ or a join predicate $I_h = A_k$, where A_k is any field of another invocable access pattern AP_j; in the latter case, AP_j provides bindings to some input fields of AP_i. A query Q is then said to be *feasible* when all its access patterns are invocable. In the example, Theater and Movie are immediately invocable, while Restaurant is invocable after Theater, as the Restaurant service binds its input location with the Theater's output location.

The order of invocation of access patterns also determines the type of their join. Specifically, a *pipe join directed from AP_i to AP_j* occurs whenever AP_i provides some bindings to AP_j, while a *parallel join between AP_i and AP_j* occurs when there are join conditions between output fields of the two services.

An *access plan* describes how a query Q is executed using pipe and parallel joins of its underlying services; access plans are represented as dataflow diagrams with an initial (*in*) and a final (*out*) node, and with other nodes representing services and parallel joins; sequential accesses to services based on ranking are indicated by special node symbols, and pipe joins are simply represented by arrows incoming to a service from another node.

In general, queries may have many access plans; the example query has two possible access plans, described in Fig. 11.10(a) and (b). The precedence between Theater and Restaurant causes a pipe join in both cases—from the result of the parallel join to Restaurant in case (a), and from Theater to Restaurant in case (b).

Parallel joins can be independently optimized by using the methods described in Sect. 11.2. Pipe joins require that some of the output values from one service be used

Fig. 11.10 Access plans for
the example query

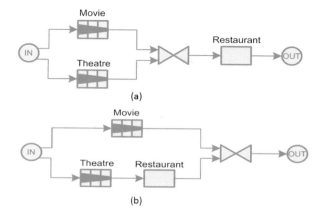

as input for calling the other one, and their top-k optimization can be performed
with specific variants of the above methods, described in [3, 247]. Strategies for
the global optimization of arbitrary queries are extremely complex, as they consist
in heuristically selecting an access plan and then assigning to each join controller
some *tokens*, each allowing a new call of the underlying services, which are either
statically defined or dynamically decided based upon the current thresholds of the
various involved joins [59].

11.4.3 Exploratory Search

Exploratory search is a process in which the user starts from a not-so-well-defined
information need and progressively discovers more both on his need and on the
available information to address it, with a mix of lookup, browsing, analysis, and
exploration. The term was first introduced by Marchionini [244] and White [349],
and will be further discussed in Chap. 14. We next describe exploratory search as
supported by the SeCo project; a limited form of exploratory search is supported by
many vertical search applications. For instance, travel planning systems such as Expedia (www.expedia.com) allow users to choose their flights and then progressively
explore other resources, e.g., by adding a hotel in the destination place, renting a car,
and booking other services or entertainment. Exploration is limited to a few options,
as users cannot explore beyond the application's boundaries.

In a broader exploratory search, users are allowed to progressively refine their
search process. They should be at least partially aware of the conceptual structure
of information, by knowing the entities that can be searched and the relationships
between them. Thus, during an exploration, users may first select a show and then relate it to several other concepts, such as the performing artist, nearby restaurants, the
transportation and parking facilities, other shows playing on the same night in town,
and so on. Technically, this means exploring an entity relationship model (such as

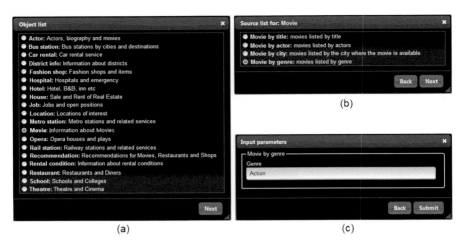

Fig. 11.11   User interface for the first exploratory step

the one shown in Fig. 11.9), with user-friendly interfaces. As the data sources describing the underlying entities are precisely described, the system is then capable of composing their data access methods by progressively building queries which express each exploratory step.

Multi-domain exploratory search provides users with ranked solutions which take into account the various ranking factors, providing at all times during interactions the shortlist of the "current best" solutions and their global ranking. This is achieved by a query and protocol which build the "current" query step by step, and then extend the "current" result step by step both structurally (the new solution includes more features) and at the instance level (the new solution includes more data items). Since items are globally ranked, it is possible to present only the top-k subset.

The SeCo system offers a user interface that allows generic explorations of a set of services modeled by an arbitrary *DD*; it is not user-friendly, but user-friendly interfaces can be built for applications by taking advantage of a preselection of domains. The system typically knows the user's current location and time, which need not be specified. The exploratory process starts by presenting a list of entities in the *DD*, as described in Fig. 11.11. After the entity selection (e.g., Movie), users are presented with options about how to access the entity (e.g., by title, actors, city, genre); in this way, they select a specific access pattern for that entity. Finally, users provide the parameters that cannot be guessed by their environments (e.g., the choice of *Action* genre). Figure 11.12 shows a further step where users, after choosing Movie and Theater, go on by choosing Restaurant, decide to access it by location and specifically next to the Theater, then choose the specific Yelp service offered by Yahoo! and finally provide the search term *Indian*. In this way, the structured query discussed in Sect. 11.4.2 can be built step by step.

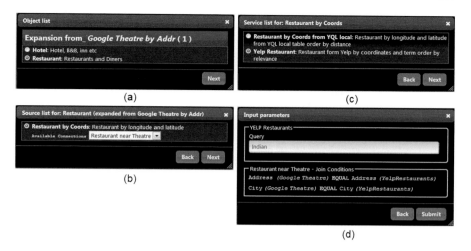

Fig. 11.12   User interface for a generic exploratory step

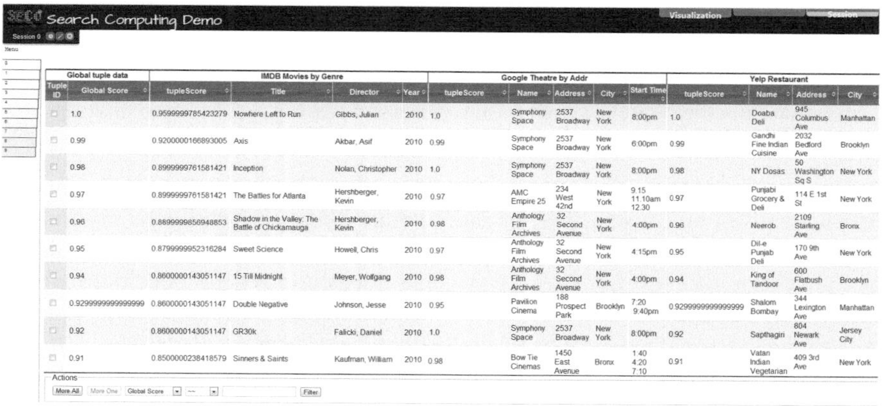

Fig. 11.13   Tabular view

11.4.4 Data Visualization

Results which are produced by a query can be visualized using several wizards. Two classic ones are tables and maps.

- The *table view* presents combinations as rows of a large table or Excel file, as shown in Fig. 11.13. The order of columns reflects the order of exploration, with the columns produced by the first step at the left of the table, followed by the columns produced at subsequent steps. Tables quickly become difficult to read with the increase of dimensions, but still are viable viewers for data analysis.

Fig. 11.14 Map view

- The *map view* presents combinations of objects which are geo-referenced on geographic maps. The information associated with those entities can be read by
clicking on the corresponding circle. A combination of multiple objects is then
a combination of positions on the map; it can be visualized by any representation that puts the objects together. In the example of Fig. 11.14, combinations are
revealed by clicking upon connecting segments.

Data representations share the same requirements. They must describe solutions
which are given by *combinations* of entity items, where each entity corresponds to
a given domain; they must show the total order of combinations which is computed
by the system. Moreover, it must be possible to perform several operations which
collectively form the *liquid query* interaction paradigm, discussed in [53]; the most
relevant operations allow users to manually filter the relevant combinations, or to ask
for more entity instances of a given specific entity or of all of them. Filtering relevant
entity instances at a given step allows users to focus their subsequent expansion steps
upon few combinations, forming the most interesting results. Additional operations
of grouping and sorting by attributes are defined only for the table views. All the
SeCo representations share a common design, with a widget (on the left side) where
the joined combinations of objects are represented (and ordered) by their global
ranking score.

11.5 Exercises

11.1 Consider the following two sources with OID-SCORE pairs:

- S_1: $\langle 3, 15\rangle$, $\langle 4, 12\rangle$, $\langle 7, 9\rangle$, $\langle 8, 7\rangle$, $\langle 1, 6\rangle$, $\langle 2, 5\rangle$, $\langle 5, 3\rangle$, $\langle 6, 1\rangle$
- S_2: $\langle 4, 10\rangle$, $\langle 7, 9\rangle$, $\langle 1, 6\rangle$, $\langle 3, 7\rangle$, $\langle 8, 4\rangle$, $\langle 2, 3\rangle$, $\langle 6, 2\rangle$, $\langle 5, 1\rangle$

Let the global score function be equal to the sum of the two scores.

1. Find the order of objects using the TA method.
2. Find the order of objects using the NRA method.
3. Compare the results.

11.2 Consider the following two sources with OID-SCORE-ATT triples:

- S_1: $\langle 1, 15, 4 \rangle$, $\langle 2, 12, 3 \rangle$, $\langle 3, 9, 2 \rangle$, $\langle 4, 7, 1 \rangle$, $\langle 5, 6, 2 \rangle$, $\langle 6, 5, 4 \rangle$
- S_2: $\langle 1, 11, 3 \rangle$, $\langle 2, 10, 2 \rangle$, $\langle 3, 9, 4 \rangle$, $\langle 4, 8, 1 \rangle$, $\langle 5, 7, 4 \rangle$, $\langle 6, 6, 3 \rangle$

Let the global score function be equal to the sum of the two scores.

1. Find the order of objects using the rank-join method.
2. Find the order of objects using the search computing method.
3. Compare the results.

11.3 Use known meta-search engines to compare hotel costs. Compare their optimal offers and make some statistics about differences in values.

11.4 Describe a new application scenario which is covered by some of the entities shown in Fig. 11.9. Describe a query in this scenario, and then express it in SQL-compatible syntax. Describe possible explorations of the search space.

11.5 Recalling Sect. 9.6, explain a possible revenue system for a search computing application involving hotels, theaters, and restaurants.

Chapter 12
Semantic Search

Abstract Semantic search seeks to answer user queries based on meanings rather than keyword matching. For instance, when searching for "museums in Milan", a traditional search engine would answer with a list of links to Web pages that contain the keywords "museums" and "Milan". However, a semantic search engine understands the intention of the user to obtain a list of museums and returns Web representations of the required real-world entities, including their semantic properties. This chapter illustrates the process adopted by semantic search engines and uses it to compare alternative semantic search solutions offered on the market as well as academic prototypes.

12.1 Understanding Semantic Search

The idea of *semantic search*, understood as searching by meanings rather than literal strings, has been investigated in the information retrieval (IR) field since the early 1980s. Enterprise semantic search engines have been around for a while, but only in recent years have start-ups like Kosmix[1] (acquired by Walmart in April 2011), Cuil,[2] Hakia,[3] and Powerset[4] (bought by Microsoft in July 2008 and now part of Bing)[5] showed that semantic search engines were ready to take their first steps on the Web. Specifically, products like WolframAlpha[6] and Google Knowledge Graph[7] addressed the usability challenge that formerly spoiled the user's experience when interacting with a semantic search engine [338]. They finally made semantics handy.

Let us consider, as a running example, the query "museums in Milan". Since the introduction of Knowledge Graph the answer of Google to this query is no longer

---

[1] http://www.kosmix.com/.

[2] http://www.cuil.com.

[3] http://www.hakia.com.

[4] http://www.powerset.com/.

[5] http://www.bing.com.

[6] http://www.wolframalpha.com/.

[7] http://www.google.com/insidesearch/features/search/knowledge.html.

S. Ceri et al., *Web Information Retrieval*, Data-Centric Systems and Applications,
DOI 10.1007/978-3-642-39314-3_12, © Springer-Verlag Berlin Heidelberg 2013

Fig. 12.1 A screen shot shows the results provided by Google Knowledge Graph when asked for "museums in Milan"

only a list of links to pages that contain the keywords "museums" and "Milan".[8]
As illustrated in Fig. 12.1, the answer also includes a top bar and a map reporting the *museums housed in Milano*. The list includes *Pinacoteca di Brera*[9]—an art
gallery—and *Museo Leonardo da Vinci*[10]—a science and technology museum. The
traditional answers would have not returned either http://www.brera.beniculturali.it/
or http://www.museoscienza.org/ since neither of the two pages contains the
searched keywords. Note that http://www.brera.beniculturali.it/ is an "art gallery"
(a "pinacoteca" in Italian) and it does not (syntactically) match the keyword "museum", but an art gallery is interpreted as a particular kind of museum and therefore
included in the answer.

In order to allow the reader to understand the key concepts behind the existing
approaches and to compare them, this chapter proposes the generic semantic search
process[11] illustrated in Fig. 12.2.

The process is divided horizontally in two parts. The user performs the steps in
the upper part, whereas the system carries out those in the lower part. In general, the
process starts when a user issues a query. The system translates the query from the
external representation (e.g., a set of keywords typed by the user) to the internal representation (e.g., a structured query). Part of the user information needs may become
lost in translation or different interpretations may exist; thus, the system presents to
the user the query as it was understood by the system. The system matches the internal query representation against the system resources, determines a ranked results

---

[8]The keyword "in" would have been considered a stop word and removed.

[9]http://www.brera.beniculturali.it/.

[10]http://www.museoscienza.org/.

[11]The semantic search process presented in this chapter is an adaptation of the one presented
in [330] with concepts from [127] and the experience gathered by the author in realizing Squiggle [81].

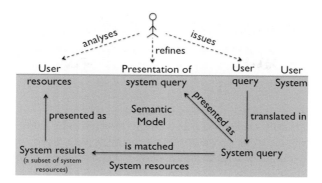

Fig. 12.2 The generic model of the semantic search process used throughout the chapter to illustrate alternative approaches

set, and presents part of it to the user. The user can then analyze the results and refine them, by choosing a different interpretation or by adding constraints.

As shown in Fig. 12.2, the *semantic model* is the cornerstone of the semantic search process. It captures the meaning of both queries and resources. The semantic model is exploited in all individual steps of the process from query construction to matching of resources, presentation of results, and refinement of the query. In particular, the semantic model has the following uses:

- It is used to translate the *query issued by the user* (e.g., "museums in Milan") into the *system query representation* used internally (e.g., a structured query that asks for all entities that are known to be museums and to be located in Milan). Sections 12.4.2 and 12.4.3 discuss the different internal query representations used and the translation techniques adopted.
- It guides the *matching* of the query against the system resources in order to retrieve a list of results. *Ranking* plays an important role in this step of the process in order to avoid information overload, and it has been a major unaddressed challenge in semantic search for decades. Section 12.5 illustrates a selection of proposed state-of-the-art semantic matching frameworks.
- It allows for an automatic *presentation of several query-related information elements*. For instance, Fig. 12.1 shows how Google Knowledge Graph answers to "museums in Milan"[12] by augmenting the usual list of links with a photo of each museum and a map showing their locations. Section 12.3.2 reviews the techniques used in current implementations with a focus on WolframAlpha.
- It is used to automatically *present the system query* to the user, who can then refine the submitted query. Facet browsing is an often employed example of an interface automatically generated from the query and the result set, which allows for adding constraints and sliding bars and selecting check boxes. Section 12.4.3 briefly discusses this topic.

---

[12]https://www.google.com/search?q=museums+in+Milan.

The process illustrated in Fig. 12.2 assumes that the semantic model is available and the resources are already present in the system in a format that allows them to be matched. As explained in Chap. 10, the recent proliferation of website publishing data on the Web following shared schemata make these assumptions reasonable. For instance, Powerset depends on Wikipedia, Google Knowledge Graphs relies on Freebase,[13] and Google's,[14] Yahoo!'s,[15] and Bing's rich snippets[16] use the data published on the Web using known schemata (e.g., schema.org).[17] However, *constructing the semantic model* and *transforming the resources* from heterogeneous formats, in which they can be published on the Web, *into the system representation* remain two important steps that must still be performed. Sections 12.6 and 12.7 elaborate these two aspects.

12.2 Semantic Model

Generally, *semantics* is about the *meaning of things*. A semantic model determines the meaning of system resources capturing the interrelationships between system resources and their interpretations, i.e., the mappings of syntactic elements in the model to real-world entities and their relations.

In the last decades, different types of semantic models have been studied in several communities:

• The *natural language processing* (*NLP*) community focused on *linguistic models* that represent the meanings at the level of words. WordNet [254],[18] a large lexical database of English, is a good example of this kind of model. It models nouns, verbs, adjectives, and adverbs as sets of cognitive synonyms (synsets in WordNet terminology), each representing a distinct meaning. For instance, the words "gallery", "art gallery", "picture gallery" are grouped in a synset described as: "a room or series of rooms where works of art are exhibited". Synsets are interlinked by means of lexical relations such as hyponyms (a "saloon" is a special kind of "art gallery", hypernyms (an "art gallery" is a special kind of "museum", and meronyms ("works of art" is part of an "art gallery"). The application of *linguistic models* to IR has been studied for a long time [144]. Several NLP companies[19] sell enterprise search engines based on these models. Powerset, which was acquired by Microsoft with the intent to be integrated within Bing, is the best

---

[13] http://www.freebase.com/.

[14] http://www.google.com/webmasters/tools/richsnippets.

[15] http://developer.yahoo.com/searchmonkey/siteowner.html.

[16] http://www.bing.com/toolbox/markup-validator.

[17] http://schema.org/.

[18] http://wordnet.princeton.edu/.

[19] http://www.crunchbase.com/tag/natural-language.

known of these approaches. However, despite the successful deployment in vertical search engines, the application of linguistic models to the Web is still under research [139].

- The *database* community focused on *conceptual models* that are concerned with meanings at the level of real-world entities denoted by words. For instance, entity-relationship diagrams [88] model objects, facts, and events of the real world as entities, and capture how entities are related to one another. A basic entity-relationship diagram for our running example contains two entities—*Museum* and *City*—and a relationship—*(isLocated)In*. Conceptual models focus on the interpretation of words in terms of the real-world entities to which the words refer; e.g., "Pinacoteca di Brera" and "Museo Leornado da Vinci" are *museums* and they *areLocatedIn* "Milano".

- The *knowledge representation* community focused on the *formalization of semantic models using logic*. For instance, entity-relationship diagrams can be given a formal semantics by an explicit mapping to first-order logic. In this way, interpretations become precise and computable. The Semantic Web community, in particular, proposes to formalize conceptual models using a family of logic languages that differ in the degree of expressivity and formality. They range from RDFS,[20] which allows one to capture simple taxonomies, to description logics (DLs) [72], which allow one to represent expressive formal models. In 2009, the OWL 2[21] standardization process at the World Wide Web Consortium (W3C) accomplished an important effort in defining three languages (namely OWL 2 profiles), each proposing a different trade-off between expressive power and efficiency of reasoning. SHOE [163],[22] the first semantic search engine built on Semantic Web technologies, KIM [200],[23] a semantic annotation, indexing, and retrieval platform commercialized by Ontotext, and TAP [150], an academic prototype experimentally deployed in the W3C website[24] as an extension of the search service to contextualize a user's query in terms of relevant W3C documents, activities, people, and services, employed semantic models whose expressivity is equivalent to the *DL-compatible fragment of RDFS*.

We next present two models that capture the semantics of the proposed example. The *minimal semantic model* is a labeled directed graph [330] where nodes represent classes, relations, attributes, and data types, and labeled edges are limited to domain and range.[25] Figure 12.3 shows the graph for the running example of Sect. 12.1. It captures museum, cities, and the relations among them.

---

[20]http://www.w3.org/TR/rdf-schema/.

[21]http://www.w3.org/TR/owl2-overview/.

[22]http://www.cs.umd.edu/projects/plus/SHOE/search/.

[23]http://www.ontotext.com/kim.

[24]http://www.w3.org/2002/05/tap/semsearch/.

[25]To be precise, nodes representing classes and data types can only appear as heads of the edges, and nodes representing relations and attributes can only appear as tails. The head of a range edge can only be a data type node if the tail is an attribute node, and it can only be a class node if the tail is a relation node. The head of a domain edge cannot be an attribute node.

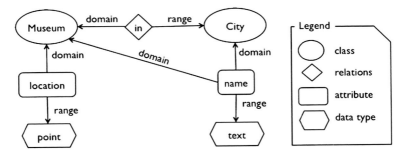

Fig. 12.3 The semantic model that captures a part of the running example proposed in Sect. 12.1 using the labeled graph proposed in [330]

This type of semantic model is at the core of AskTheWiki[26]—an academic prototype that adds semantic search to Semantic MediaWiki[27]—that was commercially exploited in the Information Workbench[28] of Fluid Operations. It is also very close to the model used for Google rich snippets, and, from what has been revealed,[29] Google is using it in Knowledge Graph. The formal semantics of this model is given by an explicit mapping to first-order logic by mapping each edge to atomic first-order formulae $p(t_1, t_2)$, where t_1, t_2 are terms and p is a binary predicate symbol. In terms of Semantic Web ontological languages, this language is a subset of the DL-compatible fragment of RDFS; it does not allow one to express hierarchies of classes (*rdfs:subClassOf*) and relations (*rdfs:subPropertyOf*).

Figure 12.4 extends the graph shown in Fig. 12.3, further capturing the notion of *building* and *place* as superclasses of museum, and *ArtGallery* as a subclass of museum.

It uses *DL-Lite$_A$* [279], the most expressive semantic model that has been successfully employed in semantic search [126], which roughly corresponds to the QL profile of OWL 2.[30] Compared to RDFS, this ontological language proposes a better notion of domain and range of properties, adds the notion of inverse property (e.g., *isLocatedIn* is the inverse of *houses*), adds the possibility to state that a property is functional (e.g., the value of the address attribute is an identifier of a building), and allow for a controlled form of negation (e.g., an *ArtGallery* is not a *ScienceAndTechnologyMuseum*).

The choice of the semantic model to use is always the result of a trade-off between conflicting requirements. T. Tran et al. in [330] identify the three following requirements:

---

[26]http://www.aifb.kit.edu/web/Spezial:ATWSpecialSearch.

[27]http://semantic-mediawiki.org/.

[28]http://iwb.fluidops.com.

[29]http://googleblog.blogspot.co.uk/2012/05/introducing-knowledge-graph-things-not.html.

[30]http://www.w3.org/TR/owl2-profiles/#OWL_2_QL.

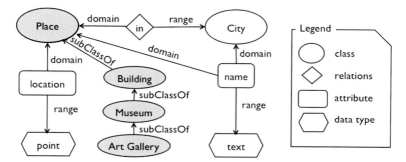

Fig. 12.4 The semantic model that entirely captures the running example proposed in Sect. 12.1 using the labeled graph proposed in [330] extended with the notion of *subclass of*

- *Tractability*: The ability to compute the reasoning tasks on the semantic model and the system resources using limited computational resources within a limited time. Tractability, of course, depends on the use case; the users of a semantic search engine for the medical domain may be willing to wait longer (seconds) than standard Web users (hundreds of milliseconds), but still they would not wait for months. The minimal graph model, presented in this section, has a very limited expressivity, but it can be practically used to support semantic search at Web scale [330]. More expressive models can be used only if appropriate techniques are employed to keep the reasoning task tractable. For instance, as explain in Sect. 12.5, when the semantic model uses $DL\text{-}Lite_A$, answers to semantic queries can be given within hundreds of milliseconds if: (a) the reasoning tasks on system resources are performed at indexing time using techniques from high-performance computing (e.g., MapReduce), and (b) reasoning at search time is limited to rewriting the issued semantic query in a set of keyword-based queries [126].
- *Generality and Extensibility*: A semantic model has to be general enough to capture a broad variety of knowledge and flexible enough to be extended with more expressive constructs. Formalizing the semantics of the model using first-order logic guarantees generality. For instance, the minimal semantic model presented in this section enables one to also capture linguistic models by interpreting words as classes and linguistic relations as predicates. Moreover, choosing to use compatible DL fragments allows for extending the minimal graph model, first with hierarchies of classes and properties and then with more expressive constructs that even include a safe form of negation.
- *Manageability*: Conceptual models are difficult to create and to maintain. In most cases manual work is needed: the more expressive the semantic model, the more work-intensive its maintenance. Effective techniques exist to automatically compute a semantic model using the minimal graph model presented in this section; machine learning techniques for *DL-Lite* are under investigation. Section 12.6 further discusses these methods.

12.3 Resources

The way resources are internally and externally represented is key to the scalability and accessibility of the search engine. This section first elaborates on the system perspective, showing two alternative solutions which are often used to internally represent resources. It then switches to the user perspective and briefly discusses the semantic-based techniques for resource visualization.

12.3.1 System Perspective

In order to answer queries in hundreds of milliseconds and to be able to scale to the Web size, resources must be internally represented in the most appropriate way. Semantic search engines often perform two different tasks: data retrieval and document retrieval. Consider our running example (search for "museums in Milan"): in the first case, the task of the search engine is to return information (e.g., name, position, address, opening hours, etc.) of museums located in Milan; in the second case, the task is to retrieve links to Web pages (or multimedia documents) that talk about museums in Milan. For instance, Fig. 12.1 shows that Google Knowledge Graph performs both types of search and presents the results to the user on the same page.

A *directed labeled graph* is normally used to represent resources for the data retrieval task. The nodes represent semantic entities and data values, the edge labels refer to the properties defined in the semantic model, and the edges assert the presence of a given property between two semantic entities (if the label of the edge refers to a relation) or between a semantic entity and a data value (if the label of the edge refers to an attribute). In some cases the label *type* is also available; edges with this label connect a semantic entity to a class defined in the semantic model. An example of a resource graph describing the entities in the running example introduced in Sect. 12.1 is illustrated in Fig. 12.5. It represents an *ArtGallery* $m1$ whose *name* is *Pinacoteca di Brera* and a *ScienceAndTechnologyMuseum* $m2$ whose *name* is *Museo Leonardo da Vinci*, both of which are located in a *City* $c1$ whose *name* is *Milan*. If the semantic model is expressive enough to allow inference, the explicit information can be augmented with inferred information. For instance, $m1$ and $m2$ are also of type *Museum*, *Building*, and *Place*. A further discussion on the role of automatic inference in semantic search follows in Sect. 12.5.

Readers who know about RDF[31] have certainly noted that the proposed graph-based model can be mapped to RDF by mapping: nodes that represent semantic entities to RDF resources, nodes that represent data values to RDF literals, and labeled edges to RDF properties.

This model (or RDF directly) was used in almost all semantic search engines that were investigated and prototyped in the 2000s for two reasons: (a) it captures not

---

[31] http://www.w3.org/RDF/.

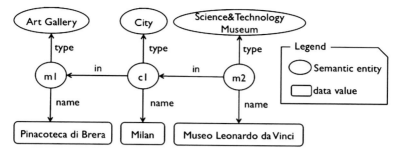

Fig. 12.5   A resource graph describing the entities in the running example introduced in Sect. 12.1

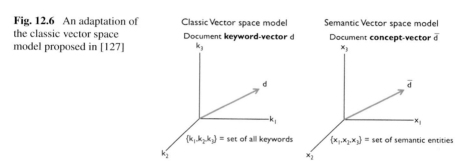

Fig. 12.6   An adaptation of the classic vector space model proposed in [127]

only graph-based data, but also tree-based (e.g., XML) and relational databases; and (b) it is supported by a large number of start-ups and major database vendors (e.g., Oracle since 11g edition[32] and DB2 since version 10 with NoSQL Graph Store[33]).

The best way to represent resources for the document retrieval is still under investigation. Web documents can be represented as nodes in the resource graph proposed above, and thus do not require any additional data structure. This is a common choice in many semantic search solutions. Alternatively [81, 127], the classic *vector space* model can be adapted to index semantic entities. As shown on the left side of Fig. 12.6, in the classic vector space model (see also Chap. 3), each document is represented as a vector where each element corresponds to a keyword, and the value of an element is the weight of the keyword, if it is present, and zero otherwise. The weight of each keyword depends on the frequency of occurrence of each keyword in each document and in the whole collection and can be computed using the term frequency-inverse document frequency (TF-IDF) algorithm. The adapted semantic vector space uses semantic entities mentioned in the document instead of keywords. For instance, if a Web page mentions the names of several museums in Milan (e.g., Pinacoteca di Brera, and Science and Technology Museum Leonardo da Vinci), the weights in the semantic vector of the semantic entities *PinacotecaDiBrera* and *MuseumLeonardoDaVinci* (mentioned in the document), *ArtGallery, ScienceAndTech-*

---

[32]http://www.oracle.com/technetwork/database/options/semantic-tech/.

[33]http://www-01.ibm.com/software/data/db2-warehouse-10/.

nologyMuseum (the explicit types of these semantic entities), and *Museum, Building*, and *Place* (the inferred types of these semantic entities) are greater than zero; weights can be computed with an adaptation of the TF-IDF algorithm that uses the frequency of semantic entities instead of the frequency of keywords. Section 12.7 provides an example in Fig. 12.15.

12.3.2 User Perspective

In early attempts to build semantic search engines, users often had to interact with the underlying semantic models and resource representation during the search process. Lack of usability was a major point of failure of semantic search until recent years [338]. While the semantics should be expressive enough to satisfy the users' information needs, the user should not be asked to deal with it. The semantic model should, instead, enable presentation of the information to the users in the most intuitive form.

Generic interfaces such as lists, tables, and trees should be proposed only when no better visualization can be found. Knowing the types of semantic entities, *fact- and entity-specific interfaces* can be generated. For instance, in the running example, museums are a special type of place that, always being described with geo-coordinates, can be visualized on a map (see the map presented by Google Knowledge Graph in the right part of Fig. 12.1). In a similar way, time series can be mapped on a timeline. For specific facts of a certain type (e.g., weather conditions), a specific presentation module may be present.

For example, Fig. 12.7 shows the results provided by WolframAlpha when asked about the "Weather in Milan".[34] Interested readers may want to browse the large set of visualization options provided by WolframAlpha[35] to understand the possibilities offered by the availability of a semantic model of the data.

12.4 Queries

As previously discussed in Sect. 12.3.2, while the inner semantic model should be expressive enough to capture the users' information needs, the user should not be asked to handle semantics. In order to be usable [338], a semantic search engine should focus on users' information needs, and should allow the user to express the information need in the most natural way, taking care of translating it (without losing too much information) in the internal representation of the query. For meaning disambiguation and query refinement, it should present to the user the information need as captured by the system query. The rest of this section presents different

---

[34]http://www.wolframalpha.com/input/?i=weather+in+Milan.

[35]http://www.wolframalpha.com/gallery.html.

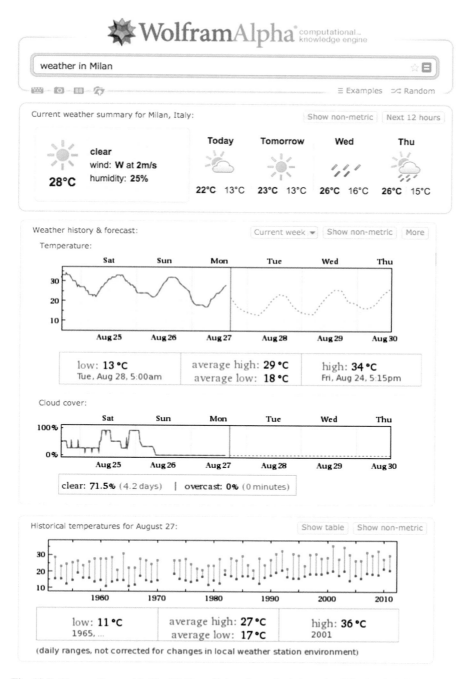

Fig. 12.7 The results provided by WolframAlpha when asked about the "Weather in Milan"

types of semantic queries, their internal representation, and the presentation to the user of the query as interpreted by the system.

12.4.1 User Perspective

Semantic search engines focus on the following kinds of information needs:

- *Entity Search*: In this kind of query the user searches for an entity, such as a document, an image, a movie, or a song. Since the search entity is somehow known, the information need can be expressed in a detailed way, e.g., "the president of the USA", "movies with Jennifer Lopez", "books of JRR Tolkien", etc. In some cases the user may even be allowed to add constraints, e.g., "the president of the USA in 1961" or "movies with Jennifer Lopez in 2004". Both Google Knowledge Graph and WolframAlpha provide answers to the three basic queries, while the latter is also able to handle constraints.
- *Fact Search*: In this kind of query the user knows about an entity and wants to be informed about given properties of that entity. For instance, both the running example query "museums in Milan" and the query "weather in Milan", whose result on WolframAlpha is illustrated in Fig. 12.7, belong to this category. Other examples of fact search queries that WolframAlpha is able to answer are: "where was Barack Obama born?", "when was 'Shall we dance' released?", and "when was the lord of the rings written?"
- *Relation Queries*: In relation queries the user want to be informed about a complex set of entities and, in particular, about the relations between the entities. Examples of such queries are: "good restaurants nearby museums in Milan", "presidents of USA born in (one of the states of) the (US) West Coast", "movies of actresses that are also singers", "fantasy books written by (University of) Oxford professors". Commercial search engines are not able to answer these kinds of queries, which are of special interest to current investigations like [54, 126, 330].

12.4.2 System Perspective

The representation of the information need expressed by a query depends on the representation of system resources, on the semantic model, and on the semantic matching framework.

If the resources are represented in a (semantic) vector space (see Fig. 12.6), the query must also be represented as a *(semantic) vector*. In approaches like [81, 127, 150], this choice also constrains whatever the user writes in the search box to be interpreted as a set of keywords, and, thus, *relation queries* cannot be answered. However, this is not always the case; for instance, in [126] the semantic model and the matching framework (based on *DL-Lite*) allow for expressing complex structured queries that are rewritten in a set of (syntactic) keyword-based queries answerable

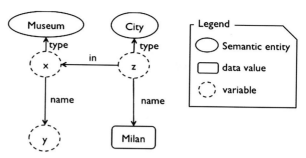

Fig. 12.8 A graph structure query that captures the running example query introduced in Sect. 12.1: "museums in Milan"

by traditional search engines. The approaches based on a (semantic) vector space also have the advantage of intrinsically providing ranked results without requiring a specific ranking algorithm to be integrated (this is discussed further in Sect. 12.5).

If the resources are represented in a graph (see Fig. 12.5), a structured query language can be employed internally and, thus, a large number of (controlled) natural language queries can be fully interpreted. This is the approach used in Google Knowledge Graph, WolframAlpha, and a wide number of academic prototypes (e.g., [127, 139, 330]).

Even if some semantic search engines (e.g., [127]) allow for internally representing queries using a relational query language such as SPARQL,[36] the best performing solutions (e.g., [126, 330]) use *conjunctive queries*.

Conjunctive queries are a fragment of first-order logic whose formulae can be constructed from atomic formulae using conjunction $\land$ and existential quantification $\exists$, avoiding disjunction $\lor$, negation $\neg$, or universal quantification $\forall$. Conjunctive queries have the form

$$(x_1, \ldots, x_k).\exists x_{k+1}, \ldots x_m.A_1 \land \cdots \land A_r$$

where the unbound variables $x_1, \ldots, x_k$ are called distinguished variables, the bound variables $x_{k+1}, \ldots, x_m$ are called undistinguished variables, and $A_1, \ldots, A_r$ are atomic formulae. For instance, the running example query "museums in Milan" can be formulated as

$$(y).\exists(x, z).name(x, y) \land type(x, Museum) \land in(x, z)$$

$$\land\ type(z, City) \land name(z, Milan)$$

that asks for the names y of semantic entities x of type *Museum* that are located in a semantic entity z of type *City* whose name is *Milan*. Figure 12.8 illustrates a graphical representation of such a query. The graphical formalism is compatible with the graph model used in Fig. 12.5 to represent the resources.

---

[36]http://www.w3.org/2009/sparql/wiki/.

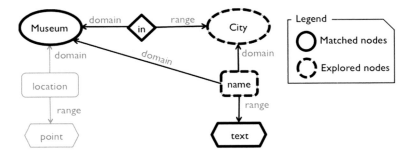

Fig. 12.9 The query space of the running example query introduced in Sect. 12.1: "museums in Milan"

12.4.3 Query Translation and Presentation

The automatic translation of a string input by the user into the query representation is both mandatory (to achieve usability) and quite hard. When multiple interpretations of the same user text are possible, the alternative interpretations could be presented to users, who then could choose the intended one. The problem has been studied for a long time in the database community in the context of natural language interfaces to databases [366], where systems like BANKS [7], DISCOVER [171], and DBXplorer [12] have focused on matching user keywords to names of columns of tables and values in rows. In the natural language community, query translation is considered as a special (hard) case of semantic text analysis (see the related work of [252] for a short survey).

In semantic search engines, the semantic model can be used to interpret the query. For instance, NAGAconjunctive queries [192] has been shown to increase the effectiveness of translation with respect to (w.r.t.) keyword-based approaches; e.g., [117] reports on NAGA outperforming BANKS. Techniques similar to the one used in NAGA are adopted in academic prototypes of semantic search engines (e.g., [127, 330]) and, most likely, are at the basis of query interpretation in WolframAlpha and Google Knowledge Graph.

These approaches, roughly speaking, break down the translation problem into three steps [330]:

- *Construction of the Query Space*: This step is similar to the one performed in keyword-based access to databases. The strings in the user input are matched against the semantic model (not the resources) as the space of possible interpretations of the input strings. The query space corresponds to: (a) the subset of the semantic model that contains the classes and properties that match the input strings (namely matched nodes), and (b) the nodes that connect the matched nodes (namely connecting nodes). Figure 12.9 illustrates the query space for the running example "museums in Milan".
- *Top-k Query Graph Exploration*: Each matched node is used as a starting point for the exploration of all distinct paths that connect them. At each step a score is

computed for each path considering: (a) the popularity of the nodes in the path (computed with an adaptation of the TF-IDF algorithm), (b) the matching score (e.g., Levenshtein distance) of the matched node, and (c) the length of the path. Step after step, the paths are merged to form query graphs that are added to a candidate list. The process continues until the upper bound score for the query graphs yet to be explored is lower than the score of the k-ranked query graph in the candidate list; i.e. no candidates can have a better score than the k-ranked result.

- *Result Presentation*: Candidate query graphs can then be presented to the user, who can assess if the interpretation provided by the system is correct or choose among alternative interpretations.

Presenting the query to the user can be particularly challenging. WolframAlpha, for instance, uses a tabular presentation. Table 12.1 (in particular, the second and third columns) shows the interpretations provided by WolframAlpha for the entity search and fact search queries presented in Sect. 12.4.1. Alternative presentations include graphs, natural language sentences, and fact-specific visualizations. Figure 12.8 is an example of a graph representing the running example query. Some systems may suggest query refinements; for instance, if one writes "museums in New York Cities" in the Google search box, it suggests to add "open on Mondays"[37] in order to filter the results based on the opening days. Facets are values assumed by a set of predicates that describe the semantic entities returned by the query; query refinement by using facets consists of selecting specific facet values or restricting their value ranges.

12.5 Semantic Matching

Matching the internal query representation depends on the system resource representation, the semantic model, and the semantic matching framework. The first two aspects have already been discussed in Sects. 12.3 and 12.2, respectively. This section focuses on three types of semantic matching frameworks, which are based on either a semantic vector space, graph pattern matching, or a hybrid model.

If resources and queries are represented in a (*semantic*) *vector*, the matching can be performed as in a classic IR system (see Fig. 12.10) by computing the similarity between the query and the documents. If all implicit semantic entities automatically inferable from the explicit ones are added to the document vector at indexing time matching, the running example query — "museums in Milan"—returns both *PinacotecaDiBrera*, which being an *ArtGallery* is also a *museum*, and *MuseoLeonardoDaVinice*, which being a *ScienceAndTechnologyMuseum* is also a *museum*. This approach is used in [81, 127]; most likely, it is also used by Google Knowledge Graph.

---

[37] https://www.google.com/search?q=museums+in+NYC+open+on+mondays.

Table 12.1 The interpretations and the results provided by WolframAlpha for the entity search and fact search queries illustrated in Sect. 12.4.1

Query	Interpretation	Alternative Interpretation	Results
the president of the USA	United States | President		Barack Obama
movies with Jennifer Lopez	movies | with | Jennifer Lopez		My Family, Money Train, ... (total: 23)
books of JRR Tolkie	J.R.R. Tolkien (author)	books	information about J.R.R. Tolkien
the president of the USA in 1961	United States | President | 1961		John F. Kennedy
movies with Jennifer Lopez in 2004	movies | with | cast member | Jennifer Lopez | and | release date | 2004		Jersey Girl, Shall We Dance
museums in Milan	museums (English word)	Milan, Lombardy	noun, a depository for ...
weather in Milan	weather | Milan, Italy	weather | Milan, United States	See Fig. 12.7
where was Barack Obama born?	Barack Obama | place of birth		Honolulu, Hawaii, United States
when was shall we dance released?	Shall We Dance (2004) | release date	Shall We Dance (1937) | release date	15/10/2004
when was the lord of the rings written?	Lord of the Rings (books) | first published		The Fellowship of the Ring, 24/07/1954 (58 years ago) The Two Towers, 11/11/1954 (57 years ago) The Return of the King, 20/10/1955 (56 years ago)

Interested readers may want to compare the results on Google Knowledge Graph of "museums in NYC"[38] and "art galleries in NYC"[39] and see that the MoMA is included in both lists. The key advantage of this approach is the possibility to obtain ranked results in a straightforward manner.

If resources and queries are internally represented as graphs, *graph pattern matching* techniques can be employed. As for query translation, the semantic model serves to construct an *answer space* [192, 330] that contains data elements satisfying the structural constraints expressed by the query (i.e., in the running example, all semantic entities of type *Museum* that are *in* at least one *City*). Figure 12.11 illustrates

---

[38]https://www.google.com/search?q=museums+in+NYC.

[39]https://www.google.com/search?q=art+galleries+in+nyc.

Fig. 12.10 Adaptation of the vector space model to semantic matching

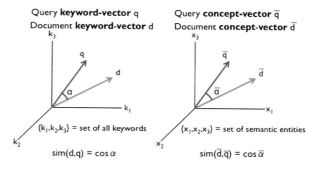

Query **keyword-vector** q
Document **keyword-vector** d

$\{k_1, k_2, k_3\}$ = set of all keywords

$sim(d,q) = \cos \alpha$

Query **concept-vector** $\bar{q}$
Document **concept-vector** $\bar{d}$

$\{x_1, x_2, x_3\}$ = set of semantic entities

$sim(\bar{d}, \bar{q}) = \cos \bar{\alpha}$

Fig. 12.11 The answer space of the running example introduced in Sect. 12.1

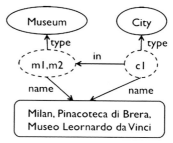

the answer space for the running example presented in Sect. 12.1. Then, semantic elements in the answer space, which also match constants, are selected and joined along the query edges. For example, in the running example, the constant "Milan" is matched and joined along the edges $\langle z\ name\ Milan \rangle$, $\langle x\ in\ z \rangle$, and $\langle x\ name\ y \rangle$, thus return $y = \{PinacotecaDiBrera, MuseoLeornardoDaVinci\}$.

Ranking has been a problematic aspect in this second approach. The popularity of semantic entities (computed with an adapted version of the TF-IDF algorithm) and the authority of semantic resources in the semantic model (computed with an adaptation of HITS or PageRank) have been employed [155], but with limited results; intuitively, hubs in semantic models may not correspond to important concepts for users, but rather to generic resources in the model. However, some promising results are appearing; for instance, NAGA [192] provides an effective solution to the problem of ranking results to fact search queries based on the trust of the knowledge source w.r.t. given semantic resources (i.e., a source can be known as trusted when the query is about museums, but known to be of poor quality when the query is about movies).

A *hybrid* approach between the two frameworks presented above is successfully employed in [126]. In this approach, publishers of Web resources are supposed to add to the Web pages where they explicitly publish data also the implicit data that can be inferred from the published one w.r.t. the employed semantic model. In this way, a classic search engine can index in a vector space both the explicit and the implicit data. Queries expressed w.r.t. a *DL-Lite$_A$* semantic model are rewritten at search time in a set of syntactic queries to be evaluated using the vector space model.

For instance, let us consider the conjunctive query that represents the running example (see also Sect. 12.4.2):

$$(y).\exists(x, z).name(x, y) \wedge type(x, Museum) \wedge in(x, z) \wedge type(z, City)$$

$$\wedge \, name(z, Milan)$$

By applying the method presented in [126], the query can be rewritten in the following two keyword-based queries:

$$Q_1 = type \; Museum \; \text{AND} \; name \; \text{AND} \; in$$

$$Q_2 = type \; City \; \text{AND} \; \text{'}name \; Milan\text{'}$$

Q_1 asks for Web pages that talk about *Museums* that are described with at least a *name* and are related to other semantic entities using the *in* property. Q_2 asks for Web pages that describe semantic entities of type *city* whose *name* is *Milan*. The result of the original conjunctive query can be built joining the results of Q_1 and Q_2. Notably, the search engines that answer Q_1 and Q_2 will return a ranked answer; therefore, a rank composition algorithm [181] can be used to rank the answer of the original conjunctive query.

12.6 Constructing the Semantic Model

Traditionally, semantic models are constructed by knowledge engineers that interview domain experts during the design of the system. This approach is applicable to enterprise semantic search, because the semantic model can be based on well-defined, domain-specific information. The knowledge of an enterprise may evolve over time, but the evolution is slow enough to be manually handled. However, to construct and maintain in this way a semantic model of the entire Web appears impossible, because the Web is a space of dynamically evolving, generic information. Viable approaches must extract the semantic model from the structured and semistructured information published on the Web, either automatically or semiautomatically.

Bisimulation [274]—a technique for the analysis of state-based dynamic systems —has been shown [330] to be an effective technique for creating a minimal semantic model of the type described in Sect. 12.2, starting from data represented as a labeled directed graph (see Sect. 12.3.1). In theoretical computer science, two dynamic systems are *bisimilar* if they match each other's state transitions, and thus an observer cannot distinguish them. Extending this concept to a labeled directed graph, two nodes are bisimilar if they cannot be distinguished by looking at their *neighborhood of edge labels*, i.e., the same structural pattern is found around them in the data graph. The algorithm employed in [330] starts partitioning the entire graph into two subgraphs, each containing bisimilar nodes, and recursively splits each sub-graph until no more splitting is possible. Each graph of bisimilar nodes obtained in this way is considered as a node representing a class or a data type in

Museum of Modern Art

From Wikipedia, the free encyclopedia
(Redirected from MoMA)

Coordinates: 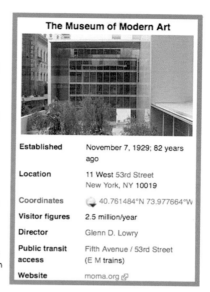 40.761484°N 73.977664°W

The Museum of Modern Art (MoMA) is an art museum in Midtown Manhattan in New York City, on 53rd Street, between Fifth and Sixth Avenues. It has been important in developing and collecting modernist art, and is often identified as the most influential museum of modern art in the world.[1] The museum's collection offers an unparalleled overview of modern and contemporary art,[2] including works of architecture and design, drawings, painting, sculpture, photography, prints, illustrated books and artist's books, film, and electronic media.

MoMA's library and archives hold over 300,000 books, artist books, and periodicals, as well as individual files on more than 70,000 artists. The archives contain primary source material related to the history of modern and contemporary art. It also houses a restaurant, The Modern, run by Alsace-born chef Gabriel Kreuther.[3]

The Museum of Modern Art	
Established	November 7, 1929; 82 years ago
Location	11 West 53rd Street New York, NY 10019
Coordinates	40.761484°N 73.977664°W
Visitor figures	2.5 million/year
Director	Glenn D. Lowry
Public transit access	Fifth Avenue / 53rd Street (E M trains)
Website	moma.org

Fig. 12.12 The infobox of the Wikipedia page about the MoMA museum contains information (e.g., when it was established, where it is located, how to get there by public transport, etc.) that can be extracted automatically

the semantic model (e.g., *City*, *Museum*, and *Text* in the running example illustrated in Fig. 12.3). A node representing a relation (e.g., *in* in the running example) is then introduced in the semantic model if there is at least a pair of nodes in the data graph ($m1$ and $c1$) belonging to two classes in the semantic model (*Museum* and *City*) such that the data graph has an edge between them.[40] A node representing an attribute (e.g., *name* in the running example) can be added, in a similar way, to the semantic model if there is at least a pair of nodes in the data graph, one belonging to a class (e.g., $c1$) and one to a data type (e.g., *Milan*).

As discussed in Chap. 10, structured data are starting to be available on the Web, but a large majority of Web data is still published in semistructured form. For instance, the upper right corner of Wikipedia pages often displays a semistructured *infobox* containing pairs of properties and values. From the infobox of the Wikipedia page about the MoMA museum,[41] shown in Fig. 12.12, readers can learn when MoMA was established, where it is located, how many people visit it in a year, and how to get there by public transport, etc. DBpedia [49], YAGO [323], and Google

---

[40]The direction of the edge in the data graph is used to add to the semantic model the edges labeled with *domain* and *range*.

[41]http://en.wikipedia.org/wiki/MoMA.

Fig. 12.13 Part of the information about the MoMA museum that DBpedia semi-automatically extracted from Wikipedia

Knowledge Graph are semantic models which were *semiautomatically extracted* from semistructured information published on the Web.

As reported on the DBpedia[42] website, the current English version describes 3.77 million things, out of which 2.35 million are classified in a consistent ontology,[43] including 764,000 persons, 573,000 places, 333,000 creative works (including 112,000 music albums, 72,000 films, and 18,000 video games), 192,000 organizations (including 45,000 companies and 42,000 educational institutions), 202,000 species, and 5,500 diseases. Localized versions of DBpedia are also available; the full DBpedia dataset describes 10.3 million things in up to 111 different languages. The dataset consists of 1.89 billion pieces of information (the unit of information is an RDF triple, i.e., two nodes and a labeled edge of the resource graph model described in Sect. 12.3.1). Figure 12.13 shows part of the facts extracted from

---

[42]http://dbpedia.org/.

[43]http://mappings.dbpedia.org/server/ontology/classes/.

Fig. 12.14 The class Museum as manually created by DBpedia authors from the most commonly used infobox templates within the English edition of Wikipedia

Museum (Show in class hierarchy)

Label (en):   museum
Label (fr):   musée
Label (ko):   박물관
Label (ja):   博物館
Super classes: Building
Properties on *Museum*:

Name	Label	Domain	Range
address (edit)	address	Building	*xsd:string*
collection (edit)	collection	Museum	*xsd:string*
curator (edit)	curator	Museum	Person
floorArea (edit)	floor area	Building	*Area*
floorCount (edit)	floor count	Building	*xsd:positiveInteger*
seatingCapacity (edit)	seating capacity	Building	*xsd:nonNegativeInteger*
structuralSystem (edit)	structural system	Building	owl:Thing
tenant (edit)	tenant	Building	Organisation

Wikipedia about the MoMA museum.[44] Interested readers may check the correspondence between Figs. 12.12 and 12.13.

DBpedia extractors not only look at the infoboxes, they also extract abstracts, inter-language links, the redirects used in Wikipedia to identify synonymous terms, and the disambiguation pages explaining the different meanings of homonyms. This extraction process is non-trivial: different communities use different infobox templates to describe the same types of things (for instance, the infobox template for the Japan cities[45] is different from the Swiss one[46]), many Wikipedia editors do not obey the templates, attribute values are expressed using a wide range of different formats and units of measurement, etc. In order to master this heterogeneity, the authors of DBpedia manually created an ontology from the most commonly used infobox templates of the English edition of Wikipedia. Figure 12.14 shows the class Museum in DBpedia ontology.[47]

YAGO[48] is an ontology similar to DBpedia that was automatically constructed from Wikipedia infoboxes and WordNet—a linguistic model previously introduced in Sect. 12.2. It is an attempt to go beyond the manual generation of DBpedia ontology. It contains around 1.7 million classes arranged in taxonomic hierarchies, obtained by combing the categories of Wikipedia with the taxonomic relations present in WordNet. 170 attributes and relations were manually created from the most frequently used attributes in Wikipedia infoboxes. It also contains 15 million facts extracted from Wikipedia infoboxes and combined with

---

[44] http://dbpedia.org/page/Museum_of_Modern_Art.

[45] http://en.wikipedia.org/wiki/Template:Infobox_city_Japan.

[46] http://en.wikipedia.org/wiki/Template:Infobox_Swiss_town.

[47] http://mappings.dbpedia.org/server/ontology/classes/Museum.

[48] http://www.mpii.de/yago.

those present in WordNet. An important aspect of YAGO is the quality control. The authors developed two rigorous quality control mechanisms: canonicalization and type checking. Canonicalization makes each fact and each class identifier unique. Type checking eliminates individuals that do not have a class and facts that do not respect the domain and range constraints of their relation.

Google Knowledge Graph was created with methods similar to those employed in DBpedia and YAGO from public sources such as Freebase, Wikipedia, and the CIA World Factbook.[49] The current version includes around 500 million semantic entities described and connected by 3.5 billion attributes and relations. Its quality is continually enhanced both implicitly, by observing what Google users search for and what they select, and explicitly, by asking for feedback from the users.

12.7 Semantic Resources Annotation

As discussed in Sect. 12.3.1, two forms of semantic search have been studied for document and data retrieval, respectively. While many site owners are publishing data on the Web enabling data search, a large majority of Web content is (and will remain) not semantically annotated. Thus, semantic annotation techniques have to be employed. These techniques are largely used in enterprise semantic search engines, and a number of services are available on the Web to automatically enrich Web pages with semantic annotations (e.g., OpenCalais,[50] Zemanta,[51] and DBpedia Spotlight[52]).

Figure 12.15 exemplifies how an NLP-based semantic annotation tool handles the Wikipedia abstract of Pinacoteca di Brera, the art gallery proposed in the running example of Sect. 12.1. The process proceeds as follows:

1. a part-of-speech parser identifies proper nouns (tagged with NNP in the figure), verbs (VBZ), and prepositions (IN);
2. a dependency parser identifies dependencies between nouns, verbs, and prepositions, e.g., nominal subject dependences (nsubj), copula dependences (cop), and prepositional modifier dependences (prep_in);
3. a named entity recognition processor uses these linguistic annotations to identify sets of words that together represent known entities, e.g., "Pinacoteca di Brera" and "Brera Art Gallery" are recognized as *organizations*, "Milan" and "Italy" as *locations*;

---

[49]https://www.cia.gov/library/publications/the-world-factbook/.

[50]http://www.opencalais.com/.

[51]http://www.zemanta.com/.

[52]http://spotlight.dbpedia.org/.

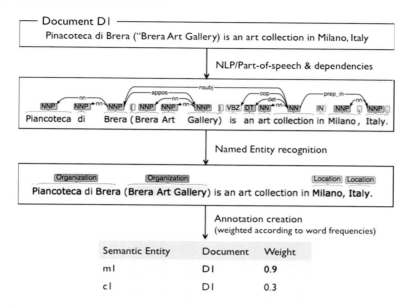

Fig. 12.15 NLP-based annotation of the Wikipedia abstract about Pinacoteca di Brera

4. an annotation creator looks up the identifiers of the recognized named entities in the semantic model and in the resource graph; it checks their frequency in the document collection, and it adds weighted annotations to the document. For instance, in the example, the annotation $m1$—*Pinacoteca di Brera* is assumed to appear less frequently than the annotation $c1$—*Milan* and, thus, it is assigned a higher weight.

In semantic search engines, the presence of a semantic model and of a resource graph can enhance NLP-based named entity recognition solution. M.A. Yosef et al. [363] showed that by using YAGO and the anchor text to Wikipedia pages, it is straightforward to compute the popularity of each pair of anchor text and semantic entity, thus obtaining a ranked list of anchor texts that are used when mentioning a semantic entity. For instance, the anchor text most frequently used when referring to "Pinacoteca di Brera" is "Brera Art Gallery", followed by "Brera Gallery" and "Brera". Moreover, a number of similarity measures can be computed using the contexts of both the anchor text in the Wikipedia page and the semantic entity in the resource graph. For each semantic entity, two bags of words can be computed: one capturing the textual surrounding of the anchor text and one capturing the entity surrounding the semantic entity. These bags of words can be used for reasoning about the coherence of recognized named entities. For example, once "you know who" has been mapped to *Lord Voldemort* or "Manchester" to *Manchester United*, it is likely that the meanings of "Harry" and "Liverpool" are *Harry Potter* and *FC Liverpool*.

12.8 Conclusions and Outlook

This chapter presents a wide range of commercial semantic search services and a selection of academic prototypes[53] that differ in the scope, in the semantic model, in the internal resource representation, in the type of allowed queries, in the internal representation of queries, in the semantic matching framework, in the ranking of results, in the presentation of results, in the presentation of queries for refinement, and in the methods used to construct the semantic model. Table 12.2 wraps up the various approaches for each of these criteria.

The availability of commercial semantic search engines witnesses the maturity reached in recent years by semantic technologies. Google Knowledge Graph and WolframAlpha are already revolutionizing Web search, by offering to Web users new services that were previously only available in enterprise search engines. Semantic entity and fact search are limited only by the availability of structured data published on the Web. However, several challenges still need to be addressed to bring semantic search to its full potential:

- Scalability to extremely large data sets or document collections is still an issue, especially for relation search queries that suffer from combinatorial explosion [54].
- An appropriate approach to rank semantic search results is available only for entity search (e.g., in [126, 127, 330]). Promising results are appearing for fact search (e.g. in [192]), but the general problem of ranking semantic search results is still under investigation.
- Automatically building semantic models has been shown to be practically possible if data are complete and correct (e.g., in [330]), or if a large amount of information is manually entered in semistructured forms (e.g., Wikipedia infoboxes) so as to enable its extraction (e.g., [49, 323]). However,

  - Methods to automatically build expressive semantic models (e.g., in *DL-Lite*) are still under investigation,
  - The lack of completeness and the presence of noise in the resource graph call for the usage of inductive reasoning techniques in constructive the semantic model, but no tractable solution has been identified, and
  - The effectiveness of information extraction techniques at Web scale (i.e., beyond Wikipedia) has not yet been proved.

- Canonicalization of semantic entities identifiers, even if possible when constructing a semantic model from a single source [323], is an open problem at Web scale and the distributed publishing of data on the Web will likely exacerbate it.
- Effective visual interfaces to present search results and allow for query refinement require more investigation.

---

[53]For a broader coverage of the academic works, interested readers are referred to two recent surveys on semantic search [125, 338].

Table 12.2 A comparison of the different approaches to semantic search

Criterion	Approach
Scope	Document search
	Data search
Semantic model	Directed labeled graph (DLG)
	DLG + class and property hierarchies (DL-compatible fragment of RDFS)
	DL-Lite
Internal resource representation	Semantic vector space
	Directed labeled graph (RDF)
Type of allowed queries	Entity search
	Fact search
	Relation search
Internal representation of queries	Semantic vector space
	Conjunctive queries
	Relational query language (SPARQL)
Semantic matching framework	Semantic vector space
	Graph matching
	Query evaluation
Ranking of results	Adapted TF-IDF algorithm
	Adapted HIT or PageRank algorithm
	Rank composition
Presentation of results	Generic presentation interfaces (e.g., tables, trees, lists)
	Fact-specific (e.g., maps, timelines)
	Entity-specific (e.g., weather in Milan)
Presentation of queries for refinement	Keywords
	Tables
	(Controlled) natural language
Construction of the semantic model	Manual
	From data (e.g., bisimulation)
	Manual for the schema and semiautomatic extraction from semistructured content for the resource graph
	Semiautomatic for both schema and resource graph from semistructured content
Semantic annotation of document	NLP-based
	Based on contextual semantic information

12.9 Exercises

12.1 Why is the semantic model the cornerstone of the semantic search process? Which roles does it play?

12.2 In a minimal semantic model represented as a labeled directed graph, what do nodes and edges represent?

12.3 Draw a resource graph in the graphical syntax used in Fig. 12.5 that captures the information contained in the following sentence: Colosseo and San Pietro are two historical sites in Rome.

12.4 What is the difference between entity search, fact search, and relation queries?

12.5 Can entity and fact search queries be answered by a search engine that uses a semantic vector space to model resources? Why?

12.6 Can relation queries be answered by a search engine that uses a semantic vector space to model resources? Why?

Chapter 13
Multimedia Search

Abstract The Web is progressively becoming a multimedia content delivery platform. This trend poses severe challenges to the information retrieval theories, techniques, and tools. This chapter defines the problem of multimedia information
retrieval with its challenges and application areas, overviews its major technical
issues, proposes a reference architecture unifying the aspects of content processing and querying, exemplifies a next-generation platform for multimedia search,
and concludes by showing the close ties between multi-domain search and multi-
modal/multimedia search.

13.1 Motivations and Challenges of Multimedia Search

The growth of digital content has reached impressive rates in the last decade, fueled
by the advent of the "Web 2.0" and the emergence of user-generated content. At the
same time, the convergence of the fixed-network Web, mobile access, and digital
television has boosted the production and consumption of audio–visual materials,
making the Web a truly multimedia platform.

This trend challenges search as we know it today, due to the more complex nature
of multimedia with respect to text in all the phases of the search process: from
the expression of the user's information need to the indexing of content and the
processing of queries by search engines.

This chapter gives a concise overview of multimedia information retrieval (MIR),
the longstanding discipline at the base of audio–visual search engines.

13.1.1 Requirements and Applications

"Finding the title and author of a song recorded with one's mobile in a crowded
disco"; "Locating news clips containing interviews to President Obama and accessing the exact point where the Health Insurance Reform is discussed"; "Finding a
song matching in mood the images to be placed in a slideshow". These are only a
few examples of what MIR is about: satisfying a user's information need that spans
across multiple media, which can itself be expressed using more than one medium.
The requirements of MIR application go beyond text IR in a variety of ways [224]:

- *Opacity of Content*: Whereas in text IR the query and the content use the same medium, MIR content is opaque, in the sense that the knowledge necessary to verify if an item is relevant to a user's query is deeply embedded in it and must be extracted by means of a complex preprocessing task (e.g., extracting speech text from a video).
- *Query Formulation Paradigm*: The user's information need can be formulated not only by means of keywords, as in traditional search engines, but also by analogy, e.g., by providing a sample of content "similar" to what the user is searching for.
- *Relevance Computation*: In text search, the relevance of documents to the user's query is computed as the similarity degree between the vectors of *words* appearing in the document and in the query (modulo lexical transformations). In MIR, the comparison must be done on a wide variety of features, characteristic not only of the specific medium in which the content and the query are expressed, but even of the application domain (e.g., two audio files can be deemed similar in a music similarity search context, but dissimilar in a topic-based search application).

MIR applications requirements have been extensively addressed in the last three decades, both in the industrial and academic fields. As a consequence, MIR is now a consolidated discipline, adopted in a wide variety of domains [248], including:

- architecture, real estate, and interior design (searching for ideas);
- broadcast media selection (radio channels [162], TV channels);
- cultural services (history museums [336], art galleries, etc.);
- digital libraries (image catalogs [146], musical dictionaries, biomedical imaging catalogs [33], film, video, and radio archives);
- e-commerce (personalized advertising, on-line catalogs [122], directories of e-shops);
- education (repositories of multimedia courses, multimedia search for support material);
- home entertainment (systems for the management of personal multimedia collections [146], including home video editing [8], searching for a game, karaoke);
- investigation services (e.g., human characteristics recognition [19], forensics [20]);
- journalism (searching speeches of a certain politician [145] using his name, his voice, or his face [19]);
- multimedia directory services (yellow pages, tourist information, geographical information systems);
- multimedia editing (personalized electronic news service, media authoring [226]);
- remote sensing (cartography, ecology [364] natural resources management);
- social uses (dating services, podcasts);
- surveillance (traffic control, surface transportation, nondestructive testing in hostile environments).

13.1.2 Challenges

Multimedia search engines and their applications operate on a very heterogeneous spectrum of content, ranging from homemade content created by users to high-value premium content, like feature film video. The quality of content largely determines the kind of processing that is possible for extracting information and the kind of queries that can be answered.

Challenge 1: Content Acquisition   In text search engines, content comes either from a closed collection (a digital library) or is crawled from the open Web. In MIR, content is acquired from many sources and in multiple ways:

- from media-production devices (scanners, digital cameras, smartphones, etc.);
- by crawling the Web or local repositories;
- by the user's contribution;
- by syndicated contribution from content aggregators;
- via broadcast capture (from air/cable/satellite broadcast, IPTV, Internet TV multicast, etc.).

The most important factor affecting what can be done with multimedia assets (apart from their editorial value) is their intrinsic quality (e.g., the definition of an image or the encoding format of a video) and the quality of the *metadata* associated with them.

Metadata are textual descriptions that accompany a content element; they can range in quantity and quality, from no description (e.g., Web cam content) to multilingual data (e.g., closed captions and production metadata of motion pictures). Metadata can be found:

1. embedded within content (e.g., closed captions);
2. in surrounding Web pages or links (HTML content, link anchors, etc.);
3. in domain-specific databases (e.g., IMDb[1] for feature films);
4. in ontologies (like those listed in the DAML Ontology Library[2]).

Challenge 2: Content Normalization   In textual search engines, context is subjected to a pipeline of operations to prepare it for indexing [26]; such pre-processing includes parsing, tokenization, lemmatization, and stemming. With text, the elements of the index are of the same nature as the constitutive elements of content: words.

Multimedia content needs a more sophisticated preprocessing phase, because the elements to be indexed (called "features" or "annotations") are numerical and textual metadata that need to be extracted from raw content by means of complex algorithms.

---

[1] The Internet Movie Database: http://www.imdb.com.
[2] The DAML Ontology Library: http://www.daml.org/ontologies.

Even prior to processing content for metadata extraction, it may be necessary to submit it to a normalization step, with a twofold purpose: (1) translating the source media items represented in different native formats into a common "internal" representation format (e.g., MPEG-4 for video files), for easing the development and execution of the metadata extraction algorithms; (2) producing alternative variants of native content items, to provide freebies (free sample copies) of copyrighted elements or low resolution copies for distribution on mobile or low-bandwidth delivery channels (e.g., making a 3GP version [14] of video files for mobile phone fruition).

Challenge 3: Content Indexing   Indexes are a concise representation of the content of an object collection. Multimedia content cannot be indexed as is, but features must be extracted from it; such features must be both sufficiently representative of the content and compact to optimize storage and retrieval.

Features are traditionally grouped into two categories:

- Low-Level Features: Concisely describe physical or perceptual properties of a media element (the color or edge histogram of an image).
- High-Level Features: Domain concepts characterizing the content (extracted objects and their properties, geographical references, content categorizations, etc.).

As in text, where the retrieved keywords can be highlighted in the source document, in MIR there is also the need of locating the occurrences of matches between the user's query and the content. This requirement implies that features must be extracted from a continuous medium, and that the coordinates in space and time of their occurrence must be extracted as well (the timestamp at which a word occurs in a speech audio file, the bounding box where an object is located in an image, or both pieces of information to denote the occurrence of an object in a video).

Feature detection may even require a change of medium with respect to the original file, as in the case of speech-to-text transcription.

Challenge 4: Content Querying   Text IR starts from a user's query, formulated as a set of keywords, possibly with logical operators (AND, OR, NOT). The most popular semantics of query processing is text similarity: both the text files and the query are represented in a common logical model (the word vector model [304]), which supports some form of similarity measure (cosine similarity between word vectors).

In MIR, the expression of the user's information need allows for alternative query representation formats and matching semantics. Examples of queries can be:

- Textual: One or more keywords, to be matched against textual metadata extracted from multimedia content.
- Monomedia: A content sample in a single media (e.g., an image, a piece of audio) to be matched against an item of the same kind (e.g., query by music or image similarity, query by humming) or of a different medium (e.g., finding the movies whose soundtrack is similar to an input audio file).
- Multimedia: A content sample in a composite medium, e.g., a video file to be matched using audio similarity, image similarity, or a combination of both.

Accepting in input queries expressed by means of non-textual samples requires real-time content analysis capability, which poses severe scalability requirements on MIR architectures. Another implication of non-textual queries is the need for the MIR architecture to coordinate query processing across multiple dedicated search engines: for example, an image similarity query may be responded to by coordinating an image similarity search engine specialized in low-level features matching and a text search engine, matching high-level concepts extracted from the query (object names, music gender, etc.).

Challenge 5: Content Browsing   Unlike data queries, IR queries are approximate, and thus results are presented in order of relevance and often in a number that exceeds the user's possibility of selection. Typically, a text search engine summarizes and pages the ranked results, so that the user can quickly understand the most relevant items.

In MIR applications, understanding if a content element is relevant has additional challenges. On one side, content summarization is still an open problem [11]. For example, summarizing a video may be done in several alternative ways: by means of textual metadata, with a selection of key frames, with a preview (e.g., the first 10 seconds), or even by means of another correlated item (the free trailer of a copyrighted feature film). The interface must also permit users to quickly inspect continuous media and locate the exact point where a match has occurred. This can be done in many ways: by means of annotated time bars that permit one to jump into a video where a match occurs, with VCR-like commands, and so on. Figure 13.1 shows a portion of the user interface of the PHAROS multimedia search platform [52] for accessing video results of a query: two time bars (labeled "what we hear", "what we see") allow one to locate the instant where the matches for a query occur in the video frames and in the audio, inspect the metadata that support the match, and jump directly to the point of interest.

13.2 MIR Architecture

The architecture of a MIR system [129] can be described in very general terms as a platform for composing, verifying, and executing *search processes*, defined as complex workflows made of atomic blocks, called *search services*, as illustrated in Fig. 13.2.

At the core of the architecture, there is a *service composer*, which consists of a collection of MIR processes, i.e., specifications of search services orchestrations, executed on top of a process execution engine, a runtime environment optimized for the scalable execution of data-intensive and computation-intensive workflows. A search service is a wrapper for any software component that embodies functionality relevant to a MIR solution.

The most important categories of MIR processes are *content processes*, which have the objective of acquiring multimedia content from external sources (the user,

Fig. 13.1 Visual and aural time bars in the interface of the PHAROS search platform

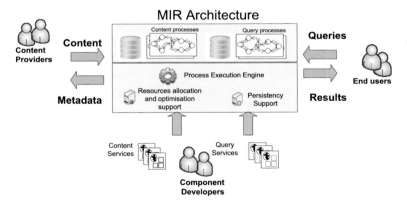

Fig. 13.2 Reference architecture of a MIR system

a video portal, a TV channel) and extracting features from it; and the *query pro-cesses*, which have the objective of acquiring a user's information need and com-puting the best possible answer to it. Accordingly, the most important categories of search services are *content services*, which embody functionality relevant to content acquisition, analysis, enrichment, and adaptation; and *query services*, which imple-ments all the steps for answering a query and computing the ranked list of results.

Examples of search services can be algorithms for extracting knowledge from media elements, transducers for modifying the encoding format of media files, query

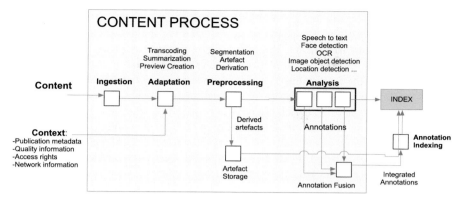

Fig. 13.3 Example of a MIR content process

disambiguation services for inferring the meaning of ambiguous information needs, or social network analysis services for inferring the preferences of a user and personalizing the results of a user's query.

13.2.1 Content Process

A content process (see the one schematized in Fig. 13.3) gathers multimedia content and prepares it for information retrieval. A MIR platform may host multiple content processes as required to elaborate content of different nature, in different domains, for different access devices, for different business goals, etc.

The input to the process is twofold:

- Multimedia content (image, audio, video).
- Information about the context where the input content is taken, which may include *publication metadata* (HTML, podcast,[3] RSS,[4] MediaRSS,[5] MPEG-7, etc.), *quality information* (encoding, user's rating, owner's ratings, classification data), access rights (digital rights management (DRM) data, licensing, user's subscriptions), and *network information* (type and capacity of the link between the MIR platform and the content source site—e.g., access can be local disk-based, or remote though a LAN/SAN, a fixed WAN, a wireless WAN, etc.).

The output of the process is the textual representation of the metadata (in the form of semantic *annotations*) that capture the knowledge extracted from the multimedia content via content processing operations (for example, textual labels that

---

[3]Podcast, http://en.wikipedia.org/wiki/Podcasting.

[4]RSS, http://en.wikipedia.org/wiki/RSS_(file_format).

[5]Media RSS, http://en.wikipedia.org/wiki/Media_RSS.

define the content of an image, the mood of a song, etc.). Sections 13.4 and 13.5, respectively, discuss the state of the art of metadata vocabularies and analysis techniques for multimedia contents. The typical steps of a content process comprise the *ingestion* of the content into the platform (via crawling, upload), the *adaptation* of the content (e.g., the transcoding to an internal format), its *preprocessing* (e.g, video segmentation, the extraction of derivatives, such as thumbnails and summaries), and finally the proper *analysis*, which is a possibly complex workflow of analysis operators, each extracting a given low-level feature of high-level annotation. Annotations derived independently can be integrated by a *fusion* step, to reinforce the confidence in the extracted knowledge, and then indexed in the *annotation indexing* step.

A content process can also dynamically adapt to the external context, in several ways:

- By analyzing the content metadata (annotations supplied manually by the user) to dynamically decide the specific analysis operators that apply to a media element (e.g., if the collection denotes indoor content, a heuristic rule may decide to skip the execution of outdoor object detection).
- By analyzing the access rights metadata, in order to decide the derived artifacts that should be extracted (if content has limited access, it may be summarized in a freebie version for preview).
- By analyzing the geographical region where the content comes from (inferring the location of the publisher may allow the process to apply better heuristic rules for detecting the language of the speech and call the proper speech-to-text transcriptor).
- By understanding the content delivery modality (a real-time stream of a live event may be indexed with a faster, even if less precise, process for reducing the time-to-search delay interval).

13.2.2 Query Process

A MIR query process (see the one schematized in Fig. 13.4) accepts in input an information need and formulates the best possible answer from the content indexed in the MIR platform, and possibly from content indexed by other external search engines or published on the Web.

The input of the query process is an information need, which can be a keyword and/or a non-textual element, such as a content sample. A collateral source of input is the query context, which expresses additional circumstances, often implicit, about the information need. Well-known examples of query context are user preferences, past user queries and their responses, access device, location, access rights, and so on.

The query is first *acquired*, which may require the multimedia part to be subjected to a content processing step for transforming the multimedia object into a tractable representation (such as a feature vector or a set of high-level annotations).

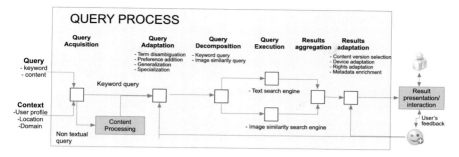

Fig. 13.4 Example of a MIR query process

Next *adaptation* is performed (for example, by disambiguating terms based on user's preferences or on context information). In this stage, the query context is used to adapt the query process, e.g., to expand the original information need of the user with additional keywords reflecting her preferences, or disambiguating a query term based on the application domain where the query process is embedded.

Query *decomposition* fragments the query into its constituents (e.g, the original keyword plus feature vectors extracted from the sample image), so as to enable query *execution* by the appropriate search engines. The results of the search engines are independent and must be aggregated to obtain a single result set that integrates results produced by all the search engines, ranked taking into account the relevance of each result in the original result sets. Prior to presentation, results can be adapted, to account for the user's preferences and access rights and for the network latency and the visualization devices.

The ultimate output of the query process is a result set, which contains information on the retrieved objects (typically content elements) that match the input query. The description of the objects in the result set can be enriched with metadata coming from sources external to the MIR platform (e.g., additional metadata on a movie taken from IMDb, or a map showing the position of the object taken from a geographic information system).

Queries are classified as *monomodal*, if they are represented in a single medium (a text keyword, a music fragment, an image), or *multimodal*, if they are represented in more than one medium (a keyword AND an image).

Queries can also be classified as *mono-domain*, if they are addressed to a single search engine (e.g., a general-purpose image search engine like Google Images[6] or a special-purpose search service like Empora[7] garments search), or *multi-domain*, if they exploit different independent search services (e.g., a face search service like Facesaerch[8] and a video search service like Blinkx [61]).

---

[6]Google Images—images.google.com.

[7]Empora Online Shop—www.empora.com.

[8]Facesaerch—www.facesaerch.com/.

13.3 MIR Metadata

The current state of the practice in content management presents a number of metadata vocabularies dealing with the description of multimedia content [137]. Many vocabularies allow the description of high-level (e.g., title, description) or low-level features (e.g., color histogram, file format), while some enable the representation of administrative information (e.g., copyright management, authors, date). In a MIR system, the adoption of a specific metadata vocabulary depends on its intended usage, especially concerning the type of content to describe. We list a set of common vocabularies currently adopted in MIR applications; the list is not intended to be complete, but rather to show how the portability, completeness, and extensibility of a metadata format can affect its usage in MIR systems.

MPEG-7 [240] represents the attempt from ISO to standardize a core set of audio–visual features and structures of descriptors and their (spatial/temporal) relationships. By trying to abstract from all the possible application domains, MPEG-7 results in an elaborate and complex standard that merges both high-level and low-level features, with multiple ways of structuring annotations. MPEG-7 is also extensible, to allow the definition of application-based or domain-based metadata.

Dublin Core [101] is a 15-element metadata vocabulary (created by domain experts in the field of digital libraries) intended to facilitate discovery of electronic resources, with no fundamental restriction on the resource type. Dublin Core holds just a small set of high-level metadata and relations (title, creator, language, etc. . . . .), but its simplicity makes it a common annotation scheme across different domains.

MXF (Material eXchange Format) [104] is an open file format, aimed at the interchange of audio–visual material, along with associated data and metadata, for various applications used in the television production chain. MXF metadata address both high-level and administrative information, like the file structure, keywords or titles, subtitles, editing notes, location, etc. Though it offers a complete vocabulary, MXF has been intended primarily as an exchange format for audio and video rather then a description format for metadata storage and retrieval.

Exchangeable Image File Format (Exif) [186] is a vocabulary adopted by digital camera manufacturers to encode high-level metadata like date and time information, the image title and description, the camera settings (e.g., exposure time, flash), the image data structure (height, width, resolution), a preview thumbnail, etc. By being embedded in picture raw content, Exif metadata is now a de facto standard for image management software. To support extensibility, Exif enables the definition of custom, manufacturer-dependent additional terms.

ID3 [179] is a tagging system that enriches audio files by embedding metadata information. ID3 includes a big set of high-level (such as title, artist, album, genre) and administrative information (e.g., the license, ownership, recording dates), but a very small set of low-level information (e.g., BPM). ID3 is a worldwide standard for audio metadata, adopted in a wide set of applications and hardware devices. Extensibility of the vocabulary is addressed in the ID3v2 version, which allows several information blocks, called frames, whose format need not be known to the software that encounters them, much in the same way as Web browsers ignore the HTML tags that they do not support.

Other examples of multimedia-specific metadata formats are SMEF[9] for video, IPTC[10] for images, and MusicXML[11] for music. In addition, several communities have created some domain-specific vocabularies, like LSCOM[12] for visual concepts, IEEE LOM[13] for educational resources, and NewsML[14] for news objects.

13.4 MIR Content Processing

Content processing is the activity performed over a content item with the aim of creating a representation suitable for indexing and retrieval purposes. The way contents are processed is application dependent, as it relates to the nature of the processed items. IR systems typically deal with textual contents. MIR systems are not an exception, as information is often represented in a textual format. Therefore, textual processing is a common activity in MIR applications, and it exploits the same standard operations for text analysis already discussed for textual IR.

MIR systems, with respect to purely textual IR systems, must also process audio, video, or images in order to produce annotations, which requires specialized operations. *Mono-annotation* analysis is defined as an analysis operation where a single combination of file type and content type (e.g., the audio track of a video file) is represented by a single annotations set. In contrast, *multi-annotation* analysis provides multiple view-points over the same content, in order to produce more descriptive annotations: for instance, the audio track of a movie can be analyzed first to identify the speaking actors and then to segment it according to speakers' turns.

Multiple annotations can be considered separately, as independent descriptions of the analyzed content, or jointly, in order, for instance, to raise the overall confidence on the produced metadata. In the former case, the annotations associated with the managed contents are defined as *monomodal*; otherwise, we talk about *multimodal* annotations. Using the example of the movie file, the fact that in a single scene both the face and the voice of a person are identified as belonging to an actor "X" can be considered as a correlated event, in order to describe the scene as "scene where actor X appears" with a high confidence. Multimodality is typically achieved by means of annotation fusion techniques: media processing operations are probabilistic processes, where the result is characterized by a confidence value; multiple features extracted from media data can be fused to yield more robust classification detection [242]. For instance, multiple content segmentation techniques (e.g., shot detection and speaker's turn segmentation) can be combined in order to achieve better video

---

[9]http://www.bbc.co.uk/guidelines/smef/.

[10]International Press Telecommunications Council: http://www.iptc.org/IPTC4XMP/.

[11]Recordare: http://www.recordare.com/xml.html.

[12]LSCOM Lexicon Definitions and Annotations, http://www.ee.columbia.edu/ln/dvmm/lscom/.

[13]Learning Object Metadata, http://ltsc.ieee.org/wg12/.

[14]International Press Telecommunications Council, http://www.iptc.org.

splitting; voice identification and face identification techniques can be fused in order to obtain better person identification. Typically, multi-annotation and multimodality can be expensive goals to achieve; thus, their adoption must be limited in such domains where fast indexing performances should be traded with accuracy in the content descriptions.

Multimedia processing operations can be roughly classified in three macro categories: *transformation, feature extraction,* and *classification.*

- *Transformation*: To convert the format of media items. For instance, a video transformer can modify a movie file in DVD quality to a format more suitable for analysis (e.g., MPEG); likewise, an audio transcoder can transform music tracks encoded in MP3 to WAV, so as to enable subsequent analysis operations.
- *Feature Extraction*: To calculate low-level representations of media contents, i.e., feature vectors, in order to derive a compact, yet descriptive, representation of a pattern of interest [318]. Such representations can be used to enable similarity search, or as input for classification tasks. Examples of visual features are *color, texture, shape*, etc. [154]; examples of music features are *loudness*, *pitch*, *tone (brightness and bandwidth)*, *mel-filtered cepstral coefficients* [334], and more.
- *Classification*: To extract and assign conceptual labels to content elements by analyzing their raw representations; the techniques required to perform this operation are commonly known as machine learning, described in Chap. 4. For instance, an image classifier can assign image files to classes (represented as high-level annotations) expressing the subject of the pictures (mountains, city, sky, sea, people, etc.); an audio file can be analyzed in order to discriminate segments containing speech from the ones containing music.

Arbitrary combinations of transformation, feature extraction, and classification operations can result in several analysis algorithms. Table 13.1 presents a list of some typical audio, image, and analysis techniques; the list is not intended to be complete, but rather to give a glimpse of the analysis capabilities currently available for MIR systems. To provide the reader with a "hook" to the recent advancements in the respective fields, each analysis technique is referenced with a recent survey on the topic.

13.5 Research Projects and Commercial Systems

In this section we overview a number of research projects that have prototyped the architecture and techniques of a MIR solution, as well as a sample of commercial systems that enable querying multimedia content.

13.5.1 Research Projects

PHAROS (Platform for searcH of Audiovisual Resources across Online Spaces) [49] is an Integrated Project of the Sixth Framework Program (FP6) of the European

Table 13.1 Content analysis techniques in MIR systems

Audio analysis	Image analysis	Video analysis
Audio segmentation [246]: to split an audio track according to the nature of its content. For instance, a file can be segmented according to the presence of noise, music, speech, etc.	Semantic concept extraction [228]: the process of associating high-level concepts (like sky, ground, water, and buildings) to pictures	Scene detection [289]: detection of scenes in a video clip; a scene is one of the subdivisions of a play in which the setting is fixed, or that presents continuous action in one place [293]
Audio event identification [281]: to identify the presence of events like gunshots and screams in an audio track	Optical character recognition [41]: to translate images of handwritten, typewritten, or printed text into an editable text	Video text detection and segmentation [267]: to detect and segment text in videos in order to apply image OCR techniques
Music genre (mood) identification [308]: to identify the genre (rock, pop, jazz, etc.) or the mood of a song	Face recognition and identification [332, 371]: to recognize the presence of a human face in an image, possibly identifying its owner	Video summarization [40]: to create a shorter version of a video by picking important segments from the original
Speech recognition [20]: to convert words spoken in an audio file into text. Speech recognition is often associated with speaker's identification [275], that is to assign an input speech signal to one person of a known group	Object detection and identification [362]: to detect and possibly identify the presence of a known object in a picture	Shot detection [97]: detection of transitions between shots. Often shot detection is performed by means of key frame segmentation [206] algorithms that segment a video track according to the key frames produced by the compression algorithm

Community, with 12 partners from 9 European countries. PHAROS has developed an extensible platform for MIR, based on the automatic annotation of multimedia content of different nature: audio, images, and video. The PHAROS content annotation process has a plug-in architecture: the content process can be defined (with a proprietary tool) and deployed in a distributed manner, possibly incorporating external components, invoked as Web services. On top of the PHAROS platform two showcase applications, one for fixed Internet and one for mobile networks, have been prototyped.

THESEUS [327] is a German research program aimed at developing a new Internet-based infrastructure to better exploit the knowledge available on the Internet. To this end, application-oriented basic technologies and technical standards are being developed and tested. For instance, the THESEUS project created and supports the Open Source project SMILA [114] (Semantic Information Logistics Architecture), a reliable, standardized industrial strength enterprise framework for building search solutions to various kinds of information (i.e., accessing unstructured information). Since June 2008, SMILA has been an official project of the Eclipse Foundation.

ALVIS [69] is an FP6 founded open source project aiming at developing open source semantic search engines in P2P-distributed architecture. The ALVIS approach consists in designing an open source infrastructure composed of separated functioning components interacting through clearly defined interfaces. Each component achieves a specific functionality to build a search process application. For example, open source components like GATE and MALLET [249] were used to provide information extraction functionalities.

Quaero [326] is a French collaborative research and development program that aims at developing multimedia and multilingual indexing, processing, and management tools to build general public search applications on large collections of multimedia information (multilingual audio, video, text, etc.). The challenge of Quaero is to integrate search and indexing components with audio/images/video processing techniques, semantic annotation methodologies, and automatic machine translation technologies, with a specific focus on improving the quality and relevance of these later technologies and techniques.

I-SEARCH [319] The EU-funded I-SEARCH FP7 project [319] has produced a unified framework for multimodal content indexing, sharing, search and retrieval, able to handle specific types of multimedia and multimodal content (text, 2D image, sketch, video, 3D objects and audio) alongside with real-world information, which can be used as queries for retrieving correlated content of any of the aforementioned types. The project has produced the Rich Unified Content Description (RUCOD), a data representation format for various multimedia types serving as a generic multimedia content descriptor, enhanced with real-world information, and expressive and emotional descriptions, supporting the retrieval of different types of media irrespective of the query format.

13.5.2 Commercial Systems

Midomi[15] applies audio processing technologies to offer a music search engine. The interface allows users to upload voice recordings of songs and then query such music files by humming or whistling. Another audio-based application is Shazam,[16] a commercial music search engine that enables users to identify tunes using their mobile phone. They can record a sample of a few seconds from any source (even with bad sound quality), and the system returns the identified song with the necessary details: artist, title, album, etc. Similar systems for music search are also provided by BMAT.[17]

Voxalead[18] is an audio search technology demonstrator implemented by Exalead to search in TV news, radio news, and VOD programs by content. The system uses a speech-to-text transcription module and transcribes political speeches in several languages.

---

[15]http://www.midomi.com.

[16]http://www.shazam.com.

[17]http://www.bmat.com.

[18]http://voxaleadnews.labs.exalead.com/?l=en.

The field of image search technologies also appears to be mature. Google Images[19] and Microsoft Bing,[20] for instance, now offer a *"show similar images"* functionality, thus proving the scalability of content-based image search on the Web. An example of an image MIR engine dedicated to a vertical domain is Chicengine,[21] which performs search for fashion-related images, allowing users to take a snapshot of a piece of garment with a mobile phone and find similar items in e-commerce sites.

In video search, *Blinkx*[22] is a search engine supporting keyword queries on both videos and audio streams. Blinkx, like Voxalead, incorporates speech recognition to match the text query to the video or audio speech content.

13.6 Exercises

13.1 Sketch the workflow of a MIR content process that supports keyword and similarity search of clothing images in a collection of user-generated photos.

13.2 Provide an example of a multimodal query that uses: text and image, text and sound, image and sound.

13.3 Describe possible applications of the classification and clustering techniques illustrated in Chap. 4 to image retrieval.

---

[19]images.google.com.

[20]http://www.bing.com/images.

[21]http://www.chicengine.com/.

[22]http://www.blinkx.com.

Chapter 14
Search Process and Interfaces

Abstract Traditional information retrieval is based on a simple paradigm: driven by
an information need, users seek information by composing a query, using a search
system; the system, in turn, associates relevant documents to the query and returns
them to the users. However, search is more a means than an end, and the needs
behind an information seeking process are typically diverse and articulated. Thus,
interaction with search systems typically involves several steps of refinement and
exploration. Visual interfaces are designed to effectively support this iterative process, help users to understand and express their information needs, and collect the
results when the process ends. This chapter provides insight into the main theoretical models used to describe the information seeking process, and offers an overview
of the user interface components used by modern search engines.

14.1 Search Process

While there are substantial technical differences between a Web search engine like
Google and the "search" command of an iPod, they share a common trait: the user
interface for the specification of a query. The simple, yet incredibly effective single
textual input field with a search button, shown in Fig. 14.1, is the de facto standard
for visual components for every search user interface.

However, these Google-like visual components are far from being the only ones:
vertical websites (e.g., news, travel, restaurants) offer a wide variety of search functions for filtering, sorting, and more; information analysis applications include advanced data visualization widgets; retailer catalogs are equipped with several tools
for browsing and exploration. Together, these advanced search functionalities are
devised to support the user in a *search process* [209], triggered after the manifesta-
tion of an information need, that is, "the perceived need for information that leads
someone to using an information retrieval system in the first place" [315].

According to Broder [65], information needs in the Web context are reflected by
three kinds of queries:

1. *Navigational queries*, whose purpose is to retrieve a particular known item that
   the user has in mind, either because she has seen it in the past, or because she
   assumes that such an item exists.

S. Ceri et al., *Web Information Retrieval*, Data-Centric Systems and Applications,   223
DOI 10.1007/978-3-642-39314-3_14, © Springer-Verlag Berlin Heidelberg 2013

Fig. 14.1 Search box of the
Bing search engine

2. *Informational queries*, whose purpose is to acquire some available information, without having a clear idea of what specific content items will be found.
3. *Transactional queries*, whose purpose is to reach a content item so that it becomes possible to perform some Web-mediated activity, with the intent of purchasing the content item.

Navigational queries, which are typically precise and simple, require user interfaces to provide simple and intuitive means to express users' needs and to access the expected items; to this end, plain keyword-based search engines serve this need well by very quickly presenting the most relevant items.

However, informational and transactional queries typically require more advanced user interfaces, which provide a wide yet intuitive set of tools for the refinement of the user needs, while providing a means to keep track of the current status of their interaction. For these classes of queries, the user's information need may have a variable degree of precision: a query may look for concrete factual information (asking the capital of a state) or look for statistically determined, imprecise information (the typical weather during a given month, accounting for its variability). Moreover, the information need can be very simple (finding the address of a given place) or more complex (determining all the relevant features that may describe a place where one would like to live).

As the complexity and imprecision of the information need increases, users engage in a search process, by doing more than one query, refining the input keywords, trying alternative formulations, and taking note of partial intermediate results, in order to assemble the answer. This is pictorially depicted in Fig. 14.2, which includes a borderline (called the note-taking boundary) delimiting a region beyond which one-shot keyword queries are not sufficient to resolve an information need, and therefore the user resorts to note taking to compose the response to his need.

The search process can be modeled as a hierarchy of goals and tasks [187], as depicted in Fig. 14.3: at the top of the hierarchy there is a *work task*, i.e., a broad activity, typically performed in the context of a working, research, or personal environment, that requires the pursuit of several goals, organized according to one's cognitive and cultural habits. The work task is executed by means of several simple *information seeking tasks*, which in turn consist of a sequence of *information retrieval tasks*; these consist of understanding information needs, leading to query specification and extraction of results, using the traditional information retrieval approach.

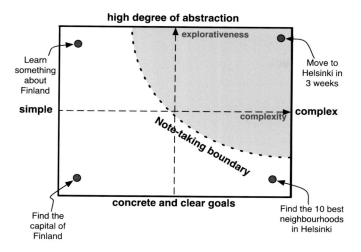

Fig. 14.2 The note-taking boundary (figure from [22])

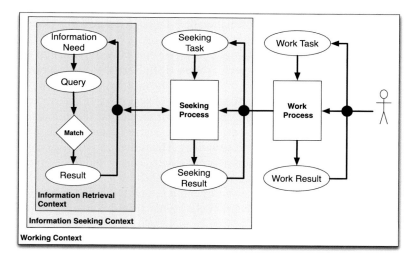

Fig. 14.3 Organization of a search process in working, seeking, and retrieval tasks (figure from [187])

14.2 Information Seeking Paradigms

Information seeking is defined as the resolution of an information need [243], typically organized as a process where users recognize their need, formulate a search, evaluate the results, reiterate the search, and, eventually, complete their task.

A very general view on the tasks and objectives of the user when exploring information is provided by the *information seeking funnel* model proposed in [298].

Fig. 14.4 The information
seeking funnel (figure from
[298], p. 2)

This model is inspired by the buying funnel or sales funnel in the commercial world,
which depicts the changing attitudes of people at different stages of the buying process, from the first interest in a product or service to the actual buy.

In the context of information seeking, the funnel model users are driven to the
bottom of the funnel, towards information consumption (see Fig. 14.4). The first
steps of the process include *wandering* (in which the user does not have an information seeking goal in mind) and *exploring* (in which the user has a general goal
but not a plan for how to achieve it). Subsequently, in the *seeking* phase, the user
clarifies the open-ended information needs that must be satisfied; finally in the *asking* phase, the user identifies an information need that corresponds to a closed-class
question.

This incremental approach is recognized by several other information seeking
models: Kuhlthau's Information Search Process (ISP) [209] depicts information as
a process of construction on the part of the individual user through a six-stage, step-
by-step model, represented in Fig. 14.5.

In [38], Bates proposed a strategic model, where she defines the different strategies and tactics a user employs in a search process (e.g., refining a search, returning
to the beginning stages, beginning a new one). The proposed *berrypicking model*
(see Fig. 14.6) assumes that users jump from source to source and from search technique to search technique as a means to build a satisfactory answer to a query.

An alternative formalization is given by the *information foraging* theory [277],
which assumes that information seekers behave like animals foraging on patches.

Model of the Information Search Process

	INITIATION	SELECTION	EXPLORATION	FORMULATION	COLLECTION	PRESENTATION	ASSESSMENT
Feelings (Affective)	Uncertainty	Optimisms	Confusion Frustration Doubt	Clarity	Sense of direction/ Confidence	Satisfaction or Disappointment	Sense of accomplishment
Thoughts (Cognitive)	vague ⟶			focused	increased interest ⟶		Increased self-awareness
Actions (Physical)	seeking	relevant Exploring ⟶	information	seeking	pertinent Documenting ⟶	information	

Fig. 14.5 The ISP model (figure from [210], p. 82).

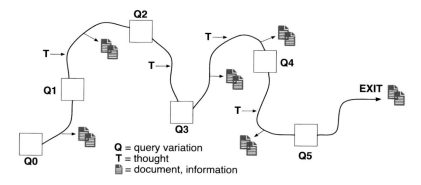

Fig. 14.6 The berrypicking model representation (figure from http://pages.gseis.ucla.edu/faculty/bates/berrypicking.html)

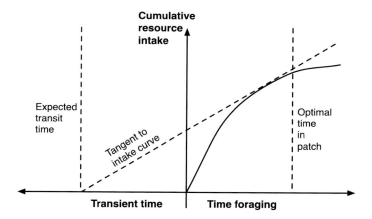

Fig. 14.7 The marginal value theorem (figure from http://en.wikipedia.org/wiki/Marginal_value_theorem)

In this case, patches are information nuggets spread all over the Web, and users move around based on patch size and patch transfer effort. They try to intake as much resources as they can, but there is an optimal time limit to be spent on a single patch for maximizing the information ingestion. The information foraging theory is in part concerned with modeling optimal strategies for foragers, that would maximize some gain (e.g., information value) function, based on the marginal value theorem, depicted in Fig. 14.7. The *transit time* refers to a new patch search, while *time foraging in patch* refers to time spent on a single patch. The slope of the tangent corresponds to the optimal rate of gain.

More recently, Marchionini [244] and White [349] introduced the notion of *exploratory search*, defined as the situation in which the user starts from a not-so-well-defined information need and progressively discovers more on his need and on the information available to address it, with a mix of lookup, browsing, analysis, and exploration.

The exploratory search discipline addresses the problem of providing the user with tools and mechanisms for easily moving through an information space until he reaches the information he was looking for. This requires supporting all the stages of information acquisition, from the initial formulation of the area of interest, to the discovery of the most relevant and authoritative sources, to the establishment of relationships among the relevant information elements. All these steps can be performed in an iterative way and should support the navigation of information along semantic connections between data. Section 11.4.3 discusses exploratory search in more detail in the context of multi-domain search.

14.3 User Interfaces for Search

As the tasks supported by search engines increase in complexity, new features are cautiously introduced in the search interface of mainstream search engines, by progressively expanding the basic paradigm of keyword entry and vertical result list presentation [160, 351]. As suggested by the very slow evolution of the interfaces provided by Google and other Web search engines, there are many well-established best practices and principles in the design of search user interfaces. This section provides an overview on the main user interaction components adopted by modern search engines.

14.3.1 Query Specification

At first glance, it seems that the *search box* (shown in Fig. 14.1), a simple yet incredibly effective single textual input field used by every search engine, has not changed much in the last years. Normally, the search box is coupled to an *advanced search* form (see Fig. 14.8(a)), where expert users can access the full expressive power of the underlying search technology by, for instance, explicitly filtering result domains (e.g., `.com`, `.net`, etc.), languages, file format, and explicit materials.

Recently, more and more search systems have started to support *Query By Example* input interfaces, where, instead of providing keywords, the users can upload an exemplary document or media. This approach is typically used for multimedia search (see more about this topic in Sect. 13.1.2), but, after several decades of investigations and adoptions in niche systems, it found application in mainstream Web search engines such as Google (see Fig. 14.8(b)).

Beyond being a purely input component, the search box offers several features aimed at supporting users in their understanding of the current state of their search process. Being visible all the time, the search box allows users to correlate the returned set of results with the current information need specification [37], expressed

YAHOO!® SEARCH _____ Yahoo! - Search Home - Help

Advanced Web Search

You can use the options on this page to create a very specific search. Just fill in the fields   [Yahoo! Search]
you need for your current search.

Show results with	all of these words	leonardo da vinci	any part of the page ⸰
	the exact phrase		any part of the page ⸰
	any of these words		any part of the page ⸰
	none of these words		any part of the page ⸰

Tip:Use these options to look for an exact phrase or to exclude pages containing certain words. You can also limit your search to certain parts of pages.

Site/Domain   ⦿ Any domain
⦾ Only **.com** domains ⦾ Only **.edu** domains
⦾ Only **.gov** domains ⦾ Only **.org** domains

⦾ only search in this domain/site: [   ]

Tip: You can search for results in a specific website (e.g. yahoo.com) or top-level domains (e.g. .com, .org, .gov).

File Format   Only find results that are: [all formats   ⸰]

(a) An extract of Yahoo's advanced search interface (`http://search.yahoo.com/web/advanced`)

Search by image   ✕
Search Google with an image instead of text.

Paste image URL | **Upload an image** ⑦

[Choose File] No file chosen   Try dragging an image here.

(b) Google's Query By Example interface for image search. (`http://images.google.com`)

Fig. 14.8 Advanced Search components in Yahoo! and Google

with a current set of keywords, thus easing query reformulation. The *autocorrection* feature helps overcome problems related to scarcity of results, possibly due to incorrect or ill-posed queries (see Fig. 14.9(a)). *Autocompletion* presents users with suggestions about queries that might be similar to the one currently typed (see Fig. 14.9(b)); suggested queries might be selected according to global usage statistics (e.g., the search activity of all Web users, or the content of the indexed Web pages), or to the users' context, e.g., their query history. Finally, autocompletion

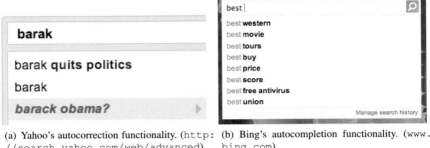

(a) Yahoo's autocorrection functionality. (http://search.yahoo.com/web/advanced)   (b) Bing's autocompletion functionality. (www.bing.com)

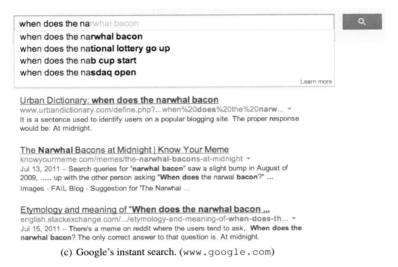

(c) Google's instant search. (www.google.com)

Fig. 14.9   Query specification suggestions

can serve as a tool to quickly answer navigational queries, e.g., queries about the weather.

Search As You Type is a search enhancement tool that allows the immediate inspection of search results matching the currently typed queries, thus helping the formulation of the query by getting instant feedback about the expected results. *Search As You Type* has recently been made popular by Google with its *Instant search* functionality (see Fig. 14.9(c)), but it has a history of earlier adoption in small-scale systems like music libraries and email searchers.

14.3.2 Result Presentation

Upon query submission, a search engine results page (SERP) is produced, containing the list of matching items, possibly ordered according to their relevance. Numer-

(a) Yahoo's search engine result page. (http://
search.yahoo.com/)

(b) Bing's search engine result page. (www.
bing.com)

(c) Google's search engine result page. (www.
google.com)

Fig. 14.10   Examples of SERP layouts in Web search engines

ous studies have investigated the impact of SERP organization and presentation on the retrieval performance of a search system, which typically features a very simple 1 to 3 column layout (see Fig. 14.10).

Recent studies [70] have identified the most attractive areas of a result page: the right part of the page is neglected, or only a short time is spent on this area, while the most considered areas depend on the task (information foraging or page recognition) and are located in the left part of the page and at the top. Almost all Web search engines rely on this model, exemplified in Fig. 14.11, typically known as the *Golden Triangle* model [160, 168, 169, 328].

Search results are typically organized in an ordered list from top to bottom, according to their relevance. For textual search engines, each result in the list includes at least (1) the *title* of the result; (2) a *snippet* of text from the matching document, aimed at providing a proof of the results relevance; and (3) a link to the indexed item.

The design of these elements has proven to be a major concern for user interface experts. For instance, a 2003 study [350] showed that snippets should highlight the matches with the search terms.

In [157], it was concluded that the visual behavior of a user engaged in a search task follows an F-shaped scanning pattern that varies according to the layout of the

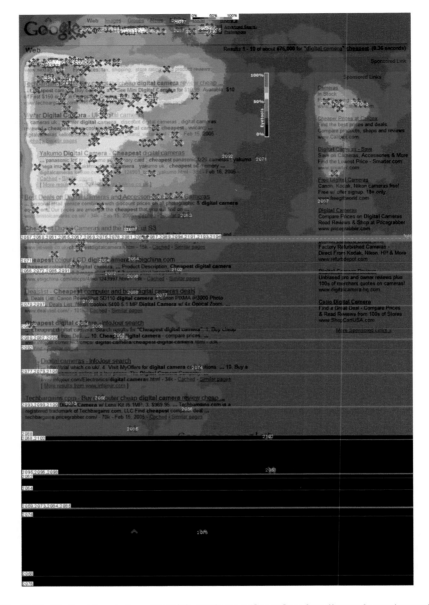

Fig. 14.11   Eyetracking study on Google's result page (figure from http://eyetools.com/research_google_eyetracking_heatmap.html)

different content types. For instance, the presence of a thumbnail changes the way in which the page is scanned; in particular, the scanning starts from the image and then the natural tendency of the eye seems to go down. The work [168] provides

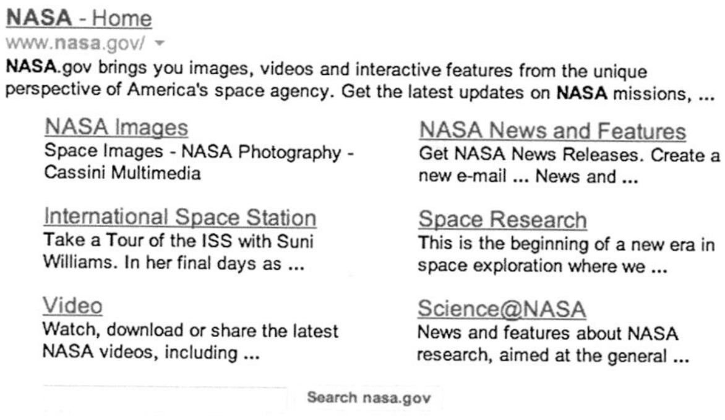

Fig. 14.12 An example of *deep linking* and *search within a site* functionality on Google's result page (http://eyetools.com/ research_google_ eyetracking_heatmap.html)

NASA - Home
www.nasa.gov/ ▾
NASA.gov brings you images, videos and interactive features from the unique perspective of America's space agency. Get the latest updates on **NASA** missions, ...

NASA Images
Space Images - NASA Photography - Cassini Multimedia

NASA News and Features
Get NASA News Releases. Create a new e-mail ... News and ...

International Space Station
Take a Tour of the ISS with Suni Williams. In her final days as ...

Space Research
This is the beginning of a new era in space exploration where we ...

Video
Watch, download or share the latest NASA videos, including ...

Science@NASA
News and features about NASA research, aimed at the general ...

Search nasa.gov

quantitative data which highlight the changes in the behavior of the users (in terms of first fixation point, scan activity, and clicks), according to the user task and to the type of results (top sponsored links, side sponsored advertisement, organic results, etc.).

Often snippets are enriched to facilitate the user's task by surfacing possibly relevant information. When the result points to a popular Web resource (e.g., the home page of a popular website), deep links allow searchers to shortcut their navigation into specific sections of the matching page, or website. The *search within a site*[1] feature provided by Google allows users to directly refine search results within a specific site. An example of application of the usage of both functionalities is shown in Fig. 14.12. As discussed in Sect. 12.4, when the indexed resource has been annotated with structured markup (e.g., RDFa), *rich snippets* allow the search results to directly show more information.

14.3.3 Faceted Search

Search results are often enriched with representative metadata, i.e., quantitative, geographical, temporal, or qualitative properties that serve the purpose of better describing the nature of the indexed item. When these metadata conform to a (possibly multidimensional) classification system, it is possible to combine text search with progressive query refinement or elaboration based on the value of the metadata; this is the essence of the *faceted search* [331] approach.

Faceted search is an advanced query mechanism where users filter a set of items by progressively selecting from only valid metadata, composed on orthogonal sets of fields (facets) [15, 119]. For instance, in the domain of wine collections, some possible facets might be the country of origin (Italy, France, Australia, etc.), the year of production (2004, 2003), the wine's color (red, white), and so on. With faceted search, users select facet values in any order, leveraging the faceted classification

---

[1] http://googleblog.blogspot.it/2008/03/search-within-site-tale-of.html.

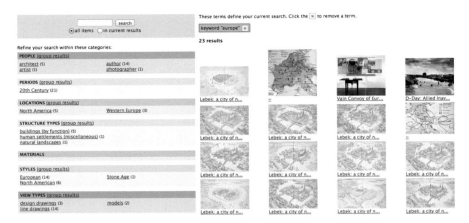

Fig. 14.13 Faceted search interface with the Flamenco navigation system (figure from http://flamenco.berkeley.edu/)

scheme to "drill-down" the result set. The advantage of this method is that null results are never achieved, while users can retrieve the desired information through a trial and error process.

Faceted search has been extensively studied, and implemented in several research prototypes, including the *Flamenco Search Interface Project* [360] (shown in Fig. 14.13), the *Relation Browser* [361], the *mSpace project* [310], and Parallax [178].

14.4 Exercises

14.1 Given the keyword queries "Washington" and "William Shatner", compare the result pages given by at least two Web search engines. Discuss the rendered layouts and type of results (maps, news, images, etc.), and provide an explanation for the observed behaviors.

14.2 Design the layout and visual components of a social network's search engine. How would it differ from the search engine of an online music streaming system?

14.3 Consider the search engine of websites like www.amazon.com or www.ebay.com, which are equipped with faceted search functionalities. Given the wealth and variety of products for sale, the number of facets and the number of facet values can be dramatically high, thus possibly hindering the information seeking process. Describe at least three possible strategies for pruning facets and facet values.

14.4 Describe the work, information seeking, and information retrieval processes that a student could adopt to provide the answer to the previous question.

Chapter 15
Human Computation and Crowdsearching

Abstract Human computation (HC) is the discipline that aims at harmonizing the
contribution of humans and computers in the resolution of complex tasks. The general principle of HC is to use computer and network architectures to organize the
distributed allocation of work to a crowd of human performers, who contribute their
skill in solving problems where algorithms fail or produce uncertain output, like
object recognition in images or text translation. HC solutions assume a variety of
forms, from crowdsourcing labor markets, to data collection or early alerting mobile
applications, to games with a purpose and crowdsearching. This chapter overviews
a few exemplary applications, discusses a conceptual framework that abstracts the
common aspects of the existing approaches, classifies the dimensions that characterize an HC solution, and highlights some open research questions and projects
addressing them.

15.1 Introduction

The Web has evolved from a publishing platform, where the interaction of users was
prevalently limited to the publication of personal content or to the access of content
created by others, to a collaborative and social tool, where users spend time online
to share information and opinions, cooperate in the execution of tasks, play games,
and participate in the collective life of communities. In the year 2011, according to
the US Digital Consumer Report by Nielsen [268], social network/blog usage and
gaming were, respectively, the first and second busiest online activities performed in
the US by fixed network users, surpassing email. In the same period, mobile usage
also registered a quantum leap (+28 %) in the share of online time spent on social
networking platforms.

The rise of gaming and social network usage sets the background for the diffusion
of a new computation paradigm, called human computation (HC) [343], applied
in business, entertainment, and science, where the online time spent by users is
harnessed to help in the cooperative solution of tasks. According to the definition of
Luis von Ahn, a pioneer in the systematic use of people in online problem solving,
human computation is *a paradigm for utilizing human processing power to solve
problems that computers cannot yet solve* [344].

S. Ceri et al., *Web Information Retrieval*, Data-Centric Systems and Applications,
DOI 10.1007/978-3-642-39314-3_15, © Springer-Verlag Berlin Heidelberg 2013

This definition is further refined by Alexander J. Quinn in [287], who distills several recent definitions from the literature into two distinctive features of an HC system:

- *The problems fit the general paradigm of computation, and as such might some-day be solvable by computers.*
- *The human participation is directed by the computational system or process.*

The common baseline of the approaches that exploit humans in computing is the intelligent partition of functionality between machines and human beings: networked machines are used for task splitting, coordination, communication, and result collection; humans participate with their intuition, decision-making power, and social links [273].

A classical example of an HC scenario is content analysis for multimedia search applications. In this domain, the goal is automatically classifying non-textual assets, such as audio, images, and video, to enable information retrieval and similarity search, for example, finding songs similar to a tune whistled by the user or images with content resembling a given picture. Recognizing the meaning of aural and visual content is one of the skills at which humans outperform machines, matured by living beings in millions of years of evolution. It is now commonly recognized that multimedia content analysis can benefit from large-scale classification performed by humans. Applications like Google Image Labeler and the system proposed by [173] submit images from a large collection to human users to receive feedback about their content and position, which can be integrated with machine-based feature extraction algorithms.

15.1.1 Background

The founding principle of HC, the structured collaboration of humans and machines in problem solving, is as old as computer science. The widespread use of the term can be traced back to the seminal work of Luis von Ahn on online games as a general incentive mechanism for encouraging human participation in problem solving [344].

HC, due to its goal of harmonizing the work of human and computer processors, is inherently a multidisciplinary topic. Figure 15.1 highlights the most relevant areas that contribute to shaping HC as a research focus.

Computer science contributes system development techniques and architectures for designing and deploying distributed systems, possibly implemented on top of heterogeneous platforms (e.g., crowdsourcing, social networks, or gaming platforms) and accessible through application programming interfaces and with multiple access devices. Besides the architectural side, human–computer interaction issues are also relevant, with a specific focus on the modeling of the user's behavior, in the design of high-quality interfaces for the execution of tasks, and in the adaptation of the user interface to different access devices.

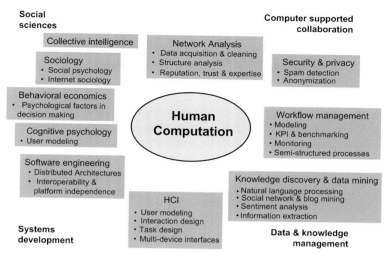

Fig. 15.1 Disciplines relevant for human computation

Data and knowledge management bring to HC the ability to extract meaning from the trails of human activities, as necessary, e.g., when the collaboration of humans is sought on such unstructured platforms as open social networks or blogs.

The organization of collective work also draws from workflow management techniques, which investigate abstract models for representing processes, key performance indicators, and quality monitoring methods, and also deal with the case of partially specified processes, where activities are conducted in part freely and in part according to some organizational constraints.

When the contribution of people is harnessed on a massive scale and on the public Internet, privacy and security research becomes relevant too. On one hand, malicious behavior detection is required, to avert individual or collective attempts at cheating with HC applications (e.g., randomly performing tasks in a crowdsourcing platform in order to gain money). On the other hand, it may be necessary to gather and process public data, while preserving the anonymity of the users who contributed them, as, e.g., in online large-scale market analysis.

When HC applications build on the social connections of people, e.g., for spreading a task or a game virally within a community, network analysis methods can be employed to understand the structure of the social ties, identify or predict the most influential members of the community, find experts on a topic, and minimize the time for completing a task. An important application of this discipline is also trust computation, which supports the selection of human performers according to their expected reliability, based on the role and activities they play in the community.

Social sciences complement the view of computer science with insight into the individual and collective cognitive and decision processes that drive the human side of HC applications. Cognitive psychology studies the mental processes that influence human behavior: perception, memory, thought, speech, and problem solving, and supports the investigation of the interactions between users and computer-based systems. Behavioral economics focuses on the specific aspect of human decision-making, a central problem in HC applications; it investigates the cognitive mechanisms that may introduce bias and non-rationality systematically into the way in which individuals judge and make decisions. Some of these mechanisms are the influence of prior knowledge, the change in preferences induced by irrelevant options, and the distorted perception of future probability based on past experience.

When HC applications exploit group behavior, the study of group dynamics as conducted in psychology, sociology, political science, and anthropology can shed light on the processes that occur within a social group or between different social groups. The methods and theories of sociology have been applied to the user groups that are constituted with the mediation of computer and network architectures, the online or virtual communities. Studies on participation, equality, and cultural diversity can be relevant to the design or to the interpretation of the results of HC applications.

Finally, HC can be seen as a specific way of harnessing the power of the so-called collective intelligence, which arises when a large number of loosely connected individuals cooperate or compete and in doing so achieve some goal that may transcend the intention and the capacity of each individual.

15.2 Applications

Under the broad umbrella of HC, several applications with different goals, architectures, and users can be recognized. In this section, we illustrate some of the most paradigmatic examples.

15.2.1 Games with a Purpose

Games with a purpose (GWAPs) focus on exploiting the billions of hours that people spend online playing computer games to solve complex problems that involve human intelligence [344]. The emphasis is on embedding a problem solving task into an enjoyable user experience, which can be conducted by individual users or by groups. GWAPs, and more generally useful applications where the user solves perceptive or cognitive problems without knowing it, address such tasks as adding descriptive tags and recognizing objects in images, checking the output of optical character recognition (OCR) for correctness, and helping with protein folding and multiple sequence alignment algorithms in molecular biology and comparative genomic research [95].

Several game design patterns have been proposed [346], which exploit different input–output templates, and the mechanics of users' involvement has begun to be modeled formally [86].

A first game template is based on the *output agreement* pattern. In this case, two randomly chosen players share a same input (typically, an image or a tune) and

(a) The ESP game   (b) The TagATune game

Fig. 15.2 Games with a purpose at www.gwap.com: output agreement (**a**) and input agreement (**b**)

are requested to produce some description of their common input: the objective is to reach an agreement as quickly as possible, by outputting the same descriptive label. Output agreement games are useful for annotating multimedia content assets, as humans are induced to produce semantic annotations that describe the input as accurately as possible. An example of output agreement games is the extra sensory perception (ESP) game [345], shown in Fig. 15.2(a).

A second game template, illustrated in Fig. 15.2(b), is based on the *input agreement* pattern, in which two players are given inputs (e.g., a tune) that are known by the game, but not by them, to be the same or different. They are requested to produce descriptions of their input, so that their partners can guess whether the inputs are the same or different. Players see only each other's outputs, and the play terminates when both players correctly determine whether they have been given the same input or not.

Both input and output agreement games assume that both players have equivalent roles, whereas the *inversion-problem* template differentiates between them: at each round, one player assumes the role of the "describer"and the other one that of the "guesser". The describer receives an input (e.g., an image, or a word) and based on it, sends suggestions to the guesser to help her make a guess that reproduces the original input. Figure 15.3(a) shows an example of an inversion-problem game, Verbosity, where the describer gives clues about a secret word to the guesser, with the aim of producing common sense descriptions of concepts. A different example of an inversion-problem game is Peekaboom [347], which is used to help detect objects in images. Here the guesser (Peek) starts out with a blank screen, while the describer (Boom) receives an image and a word related to it. The goal of the game is for the describer to progressively reveal parts of the image to the guesser, so that the latter can guess the associated word. In doing so, the game extracts a bounding box where the object associated with the image is most probably located.

GWAPs have been applied not only to the simple tasks illustrated above, but also to the massively cooperative resolution of highly complex problems, which are too expensive to address with state-of-the-art computer architectures and where human

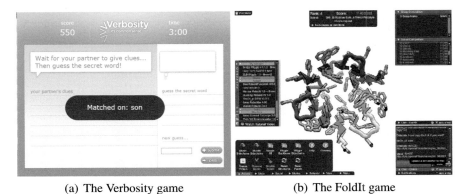

(a) The Verbosity game   (b) The FoldIt game

Fig. 15.3 Examples of games with a purpose for the inversion-problem template (**a**) and for large solution space exploration (**b**)

wisdom is used to quickly search and reduce the space of possible solutions. An exemplary case is the Foldit game [95], where crowds of online users compete and cooperate in addressing one of the hardest computational problems in biology: protein folding, that is, the prediction of biologically relevant low-energy conformations of proteins. The game has the form of a 3D puzzle, where players are given an initial structure and a set of manipulation operators that they can use to produce a folding that respects a number of physical and chemical constraints. Players can save partially folded structures, which may be taken up by other players, making the resolution effort cooperative. Several structure prediction problems have been solved successfully by Foldit players, often outperforming automated predictor tools. As an example, in 2012, Foldit players modeled in just three weeks the structure of an enzyme that had eluded researchers for a decade, which will afford new insights for the design of antiretroviral drugs [197].

15.2.2 Crowdsourcing

Crowdsourcing, a term coined by Jeff Howe in a *Wired* magazine article about the rise of new forms of online labor organizations [170], is the outsourcing of work, traditionally performed by employees, to an open community of online workers.

A first form of crowdsourcing is found in vertical markets, where a community of professional or amateur contributors is called in to supply specific products or services. An example is iStockPhoto (http://istockphoto.com), a content marketplace where photos made by both professionals and amateurs are sold at affordable prices. Other examples are found in graphic design, e.g., the 99designs design contest website (http://99designs.com) where customers can post requirements for Web page or logo creation and designers compete by submitting proposals; in fashion design, e.g., the Threadless community and marketplace for T-shirt and garment

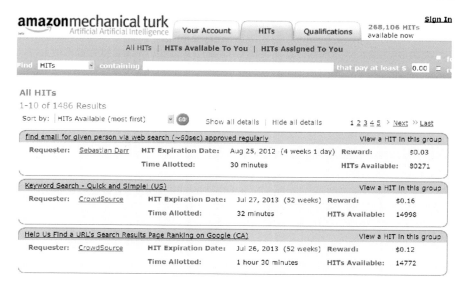

Fig. 15.4 The worker interface of the Amazon Mechanical Turk platform

design (http://www.threadless.com/); and even in highly technical and specialized services, like collaborative innovation, e.g., Innocentive (http://www.innocentive.com/), where technical challenges for product innovation are addressed to a community of problem solvers.

A different form of crowdsourcing is provided by horizontal platforms, which broker the execution of microtasks in different domains, like speech and handwriting transcription, and data collection and verification. A typical microtask brokerage platform has a Web interface that can be used by two kinds of people: work providers can enter in the system the specification of a piece of work they need (e.g., collecting addresses of businesses, classifying products by category, geo-referencing location names, etc.); work performers can enroll, declare their skills, and take up and perform a piece of work. The application manages the work life cycle: performer assignment, time and price negotiation, result submission and verification, and payment. In some cases, the application is also able to split complex tasks into microtasks that can be assigned independently [177], e.g., breaking a complex form into subforms that can be filled by different workers. In addition to the Web interface, some platforms offer application programming interfaces (APIs), whereby third parties can integrate the distributed work management functionality into their custom applications.

Examples of horizontal microtask crowdsourcing markets are Amazon Mechanical Turk and Microtask.com. Figure 15.4 shows the worker interface of Amazon Mechanical Turk. Tasks, called Human Intelligence Tasks (HITs), are packaged in groups offered as a bundle by the same requester, and groups are displayed in order of number of HITs. HITs have a descriptive title, an expiration date, a time slot for completing them, and the amount paid per solved HIT. A survey conducted in 2010 on the demographics of Amazon Mechanical Turk by Panos Ipeirotis [182] revealed that the population of workers is mainly located in the United States (46.80 %), followed by India (34.00 %), and then by the rest of the world (19.20 %), with a higher percentage of young workers; most workers spend a day or less per week on Mechanical Turk, and complete 20–100 HITs per week, which generates a weekly income of less than 20 US dollars.

15.2.3 Human Sensing and Mobilization

Crowdsourcing can take advantage of people's mobility and of the increasing diffusion of mobile terminals equipped with sensors and broadband capacity. By the end of 2011, the total number of mobile cellular subscriptions reached almost 6 billion worldwide, a penetration of 86 %, with more than 1 billion mobile broadband subscriptions, which made mobile broadband the fastest growing ICT service (+40 %) in 2011 (source: International Telecommunication union, http://www.itu.int/ITU-D/ict/statistics). More and more, mobile terminals are equipped with sensors; a US survey as of February 2012 showed that almost half of adult cell phone owners (and three-quarters of smartphone owners) use their phones to get real-time location-based information (source: Pew Research Center's Internet & American Life Project). The combination of mobile terminal diffusion, broadband, and sensors, including cameras and geo-positioning devices, provides a unique opportunity to develop large-scale crowdsourcing applications in sectors that depend on the engagement of users distributed over a territory. These applications could be used both for collecting data, when other methods are inapplicable or too costly, and for rapidly spreading information and triggering action.

Human sensing denotes the assignment of data collection tasks to a crowd [1, 74]. The focus is on the real-time collection of data, in order to realize time-critical decision support systems and emergency management. Application areas include pollution monitoring [23, 113], traffic and road condition control [34, 238], and earthquake monitoring [300].

Human sensing applications have been developed particularly in the environmental protection field, where data collection and integration is critical for decision making and human sensed observations can be used to gather a broad range of physical data, e.g., air quality in urban spaces [113], surveillance of invasive species [71], noise pollution [237], and water quality. For example, Fig. 15.5 shows Creek Watch[1] [198], a mobile and fixed Internet application whereby people can post data about watersheds rapidly, and with no instrumentation other than a standard mobile phone, like the amount of water, the rate of flow, the presence of trash and pictures of the waterway. The application design has focused both on the user interfaces, on the incentive mechanisms for engaging citizens, and on the utility of data for the scientific community that consumes them.

---

[1]www.creekwatch.org.

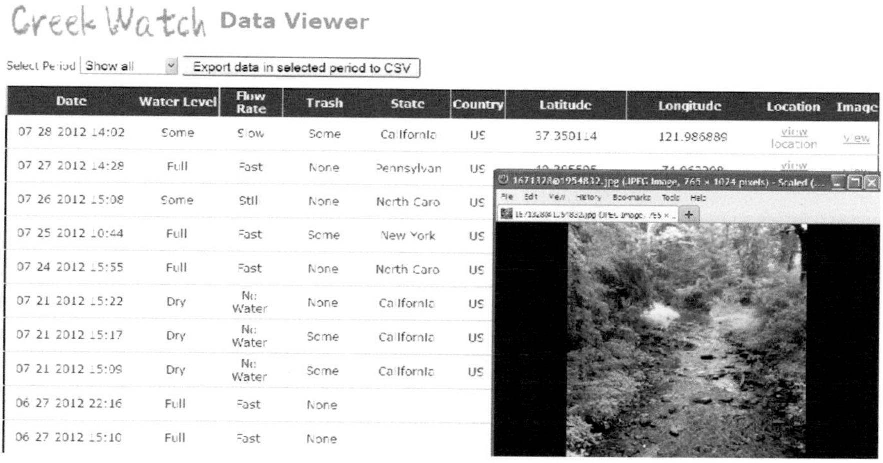

Creek Watch **Data Viewer**

Select Period [Show all ▾]   [Export data in selected period to CSV]

Date	Water Level	Flow Rate	Trash	State	Country	Latitude	Longitude	Location	Image
07 28 2012 14:02	Some	Slow	Some	California	US	37.350114	121.986888	view location	view
07 27 2012 14:28	Full	Fast	None	Pennsylvan	US			view	
07 26 2012 15:08	Some	Still	None	North Caro	US				
07 25 2012 10:44	Full	Fast	Some	New York	US				
07 24 2012 15:55	Full	Fast	None	North Caro	US				
07 21 2012 15:22	Dry	No Water	None	California	US				
07 21 2012 15:17	Dry	No Water	Some	California	US				
07 21 2012 15:09	Dry	No Water	Some	California	US				
06 27 2012 22:16	Full	Fast	None						
06 27 2012 15:10	Full	Fast	None						

Fig. 15.5 The interface of the Creek Watch data collection application

Social media have also undergone experimentation for harvesting heterogeneous and complex data, such as reporting on urban flooding events using geo-referenced tweet functionalities [230]. Using the Twitter.com microblogging service and a custom smartphone application, citizens can report events according to an existing professional controlled vocabulary (e.g., "basement flooding", "powerline down"), which is particularly useful in emergency conditions to deliver a timely response. Similar experiments of streaming human visual experience into data have been conducted in Thailand to map flooded areas and the associated damage[2] and in the Netherlands, to engage citizens in the management of emergency service, like fires.[3]

Social mobilization is an approach that goes beyond human sensing, as it aims to spread information among the population and trigger action. Its specificity is that it addresses problems with time constraints, where the efficiency of task spreading and of solution finding is essential, and exploits the social network connections among people as a vehicle for information diffusion. The DARPA Network Challenge [276] is an example of the problem and of the techniques employed to face it. The challenge required teams to determine the coordinates of ten red weather balloons placed at unknown locations in the United States. The winning team employed a novel recursive incentive mechanism that permitted them to locate all balloons in under nine hours. Applications are also found in safety critical sectors [312], like civil protection [152] and disease control [322].

---

[2] http://de21.digitalasia.chubu.ac.jp/floodmap.

[3] http://twitcident.com/.

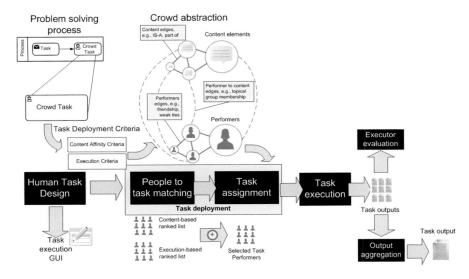

Fig. 15.6 The human computation framework

15.3 The Human Computation Framework

As discussed in Sect. 15.2, HC assumes a variety of forms, which range from paid work in an online labor market to the management of early alerts using mobile social networks. Despite such a variety of manifestations, a common framework can be identified and used to better understand the challenges in the design, usage, and evaluation of an HC solution.

Figure 15.6 conceptualizes the actors, information, and steps that are involved in HC.

15.3.1 Phases of Human Computation

By making a parallel with traditional machine-based computation, an HC framework takes in input a problem solving scenario, formalized as a *program* that can be cooperatively executed by both humans and machines. Such a program can be expressed as a workflow of elementary instructions or *tasks*, which leads to the accomplishment of the desired goal. These multiple tasks must be executed respecting precedence constraints and input–output dependencies; some of them are executed by machines (*machine tasks*) and some by a community of performers (*crowd tasks*). The community of human performers may not be completely determined a priori, and the task assignment rules are dynamic and based on the task specifications and on the characteristics of the potential members of the crowd. The HC framework produces in output some results to the problem solving goal, which may be nondeterministic, because machine tasks are solved probabilistically (as, for example, in image similarity matching) and/or because the human performers make errors, or even cheat.

A crowd task can be executed in many different ways, which depend on the specific application and on the technical platform where the contribution of humans is managed. At an abstract level, a crowd task consists of a *task user interface*, which enables the performer to fulfill the task, analogously to the program code of a machine task, and some *task execution criteria*, which can be logically subdivided into functional and non-functional criteria. Functional criteria can be regarded as the requirements on the executors' skills: in the simplest case, this can be just a descriptor of the task topic (e.g., language translation or image geo-positioning). The non-functional criteria can be regarded as constraints or desired characteristics of task execution, including a time budget for completing the work, a monetary budget for incentivizing or paying workers, bounds on the number of executors or on the number of outputs (e.g., a level of redundancy for output verification), and desired demographic properties (e.g., workers' distribution with respect to geographical position or proficiency level).

Crowd tasks are allocated to executors, analogously to the instructions of a distributed program to processors. This assignment can follow a variety of forms, but can be represented abstractly as the sequence of the steps: *task publication*, *task assignment*, and *task execution*.

- In *task publication*, the task is made known to the potential workers: this may be an explicit act, as the publication of a microtask in a crowdsourcing labor market, or implicit, when the task is embedded within an application, like a GWAP. The assignment of the task to the workers can be done in a *push mode*, when the owner of the task preselects the candidate workers; or in a *pull mode*, when the workers can bid for, or just take up, the task and start working on it.

- In *task assignment* to a selected pool of workers, the system performs a matching of people to tasks, in which the task functional and non-functional criteria are used to determine a list of potential candidate workers, possibly ranked according to their expected suitability as task executors. A worker's suitability may depend on both the task functional definition (e.g., based on the user's affinity or expertise in the topic of the task) and on non-functional criteria, such as the match between the task difficulty and user's skill level, on the role and influence of users in the network (which determines their ability to spread the task), and so on. The functional and non-functional suitability measures should be combined to obtain a globally good set of candidates.

- In *task execution*, the task is actually performed by the selected or self-proposed performers, which may result in the production of multiple outputs; these are then aggregated to form the final outcome of the crowd task. Redundancy can be exploited to cope with the uncertain quality of the worker's performance; multiple values for the same task output must be merged to obtain a final task output with a high level of confidence.

15.3.2 Human Performers

Symmetrically to the crowd task, the characteristics of the human executors may also vary greatly, depending on the application and platforms, from paid workers, to volunteers, and even to game players. Performers can be abstracted by the notion of *executor's profile*, which embodies the properties relevant to their role of task performers. The executor's profile can collect explicit information (e.g., demographic data) and information computed by analyzing the user's activity trails, for example, various indexes denoting the centrality of the user in a social network, proficiency levels determined by past activity history (e.g., game achievements), and topical affinity indicators correlating the user to the task functional specifications.

A high-level abstraction of the information space that may determine an executor's profile is a bipartite graph, where nodes denote either *performers* or *content elements*, and edges may connect performers (to denote, e.g., friendship), content elements (to denote semantic relationships), and performers to content elements (to denote, e.g., interest). The bipartite graph representation can be refined by attaching semantics to both nodes and edges: users can be associated with profile data; content elements can be summarized or classified in topics; edges between performers can express explicit friendship or implicit ties due to interactions between users; edges between content elements can express ontological knowledge, like classification, part-of, etc.; edges between users and content items can represent a specific capability (e.g., ability to review, produce, or judge content elements).

An important aspect of HC is the detection of *spammers*, i.e., users who give casual or deliberately erroneous answers. They may either perform in a superficial way (in order to maximize the number of tasks and correspondingly their monetary rewarding), or they may even act maliciously in order to interfere with the process. Thus, *spammer detection* is an important task of the HC, addressed by means of statistical evaluations of the performer's outcomes; spam detection is facilitated by the availability of a gold standard in order to determine the correctness of the HC, and thus to retrofit the system with feedbacks on the worker's performance.

15.3.3 Examples of Human Computation

Figure 15.7 shows an example of an HC program: a sequence of tasks for trademark logo detection in video collections. The problem solving goal is to receive from a user a query string denoting a brand name and to produce a report that identifies all the occurrences of logos of that brand in a given set of video files. The logo detection problem is a well-known challenge in image similarity search, where local features, e.g., the scale invariant feature transform (SIFT), are normally employed to detect the occurrences of objects based on their scale-invariant properties [231]. The computation of image similarity based on SIFT is an example of a nondeterministic machine task, and one where uncertainty may be high, because the quality of the found matches is affected by the characteristics of the input images, which

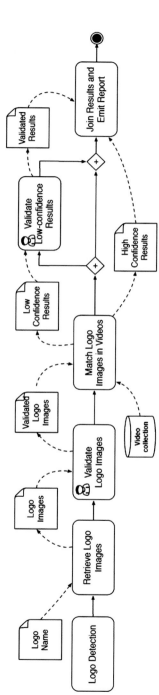

Fig. 15.7 Example of human computation program

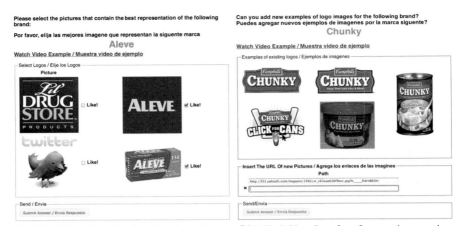

(a) A Task User Interface for choosing relevant   (b) A Task User Interface for entering new im-
images from a predefined set   ages similar to a predefined set

Fig. 15.8   Examples of task user interfaces for finding relevant and diverse images of a logo

makes it an interesting case for exploiting the contribution of humans to augment
the accuracy of the output.

The process specified in Fig. 15.7 alternates machine and human tasks: it starts
by receiving in input the brand name. The first task (*Retrieve Logo Images*) is performed by a machine executor (e.g., the Google Image search engine[4]) and processes a text-based query to retrieve representative images of the input brand name.
The output of the task may contain wrong and low-quality images, and thus the next
task (*Validate Logo Images*) is a crowd task that employs human executors to filter
the retrieved images and retain only relevant and diverse variants of the logo, so as
to enhance the accuracy of the subsequent video search phase. Then, a machine task
(*Match Images in Videos*) searches the filtered images in the collection of videos and
outputs results in the form of ⟨*videoID, frameID, matchingScore*⟩, where *matchingScore* expresses the confidence of the match found between the searched logo
and a given frame of a given video. The process continues with a second crowd task
(*Validate Low Confidence Results*), which dispatches to the crowd those matches
that have a score lower than a given threshold. Finally, the result report is constructed
(*Join Results and Emit Report* Task) by adding the high score matches found by the
algorithms and the low score matches manually validated by the human performers.

Figure 15.8 shows two examples of task user interfaces for the *Validate Logo
Images* crowd task. The functional definition of the task is simply the detection of
objects in images, whereas non-functional criteria may comprise the requirement
that each relevant images gets at least a minimum number of "likes" from workers
and that the task is completed in less than a given number of days.

---

[4]http://images.google.com/.

15.3.4 Dimensions of Human Computation Applications

The different applications of the general HC framework illustrated in Sect. 15.3 can be classified according to the dimensions of their design, which we summarize below:

- *Human Involvement*: To attain its problem solving goal, an application may require *only one individual at a time* (e.g., the reCAPTCHA system for OCR correction [348]), a *small group* of users (e.g., a multi-player GWAP) or an *open community* (e.g., people mobilization [276] and human sensing applications [113]).
- *Involved Ability*: Applications can exploit different human abilities: *perception* and/or *emotion*, as in content appraisal and user interface usability assessment, where human feedback is collected with techniques like physiological signal analysis and eye movement tracking to estimate the impact of content or of interface design [205]; *judgment*, when the discriminating power of users is exploited, as in user studies, where people are requested to judge alternative solutions [161]; *social behavior*, when cooperation and interpersonal communication are prominent, as in people mobilization [276] and social business process management, where communities of external stakeholders are involved in enterprise business processes [61].
- *Task Interface Style*: The interface of a crowd task can be based on different interaction styles. A *game* engages individuals or small groups in a challenge, where the resolution of the problem is a collateral result [86]; instead, a *task* is an explicit unit of work, with well-defined input/output and explicitly submitted to a recipient, as in crowdsourcing markets and query & answer platforms; an intermediate style is the *implicit task*, in which a user performs a useful activity without necessarily knowing it (e.g., in reCAPTCHA-enabled login forms [348]).
- *Task Control*: Task definition, assignment, and progress can be controlled centrally, as is the case in work automation platforms like Amazon Mechanical Turk. Alternatively, task activation may be managed in a distributed way, as in the DARPA Network Challenge strategy [276], where a recursive incentive mechanism was used which encouraged people to recruit effective workers so as to improve their expected success rate and reward.
- *Motivation Mechanism*: Users may engage in executing tasks for different reasons; purely for fun, as in GWAPs, for ethical reasons, as in volunteer work, or for an economical reward. Recent studies have examined the correlation between incentives, task definition policies, and the resulting quality of task execution [177].
- *Time Requirements*: HC applications may have critical time requirements that make the execution of tasks and the propagation of actions or decisions along social links important, as in the DARPA Network Challenge [276]. The design of time-critical HC applications can be addressed at the task and at the communication level. In the former case, the amount of time allotted for the execution of the task is set explicitly, so that potential performers can better self-evaluate their adequacy as problem solvers and not claim a task that they will not be able to perform

at the requested speed (this solution is adopted by most crowdsourcing applications). In the latter case, a suitable incentive mechanism is defined that favors the rapid spread of information among potential performers. As an example of this latter approach, the winners of the DARPA Network Challenge adopted a reward mechanism by which the user gained a prize not only if he spotted one balloon personally, but also if a person recruited by him spotted one or in turn recruited a third person who directly or transitively contributed to the balloon identification. This incentive fostered the rapid creation of geographically distributed teams, because each user had a quantifiable advantage at selecting from his acquaintances the most active and geographically well-positioned members to add to the team.

Figure 15.9 shows several applications of HCs and correlates them with the most relevant dimensions in their design.

15.4 Research Challenges and Projects

HC techniques and their application to various domains are the subjects of several recent research projects. We survey two projects, one that focuses on answering queries with the help of a crowd on different social networks, and one that applies HC to the problem of multimedia content processing and search.

15.4.1 The CrowdSearcher Project

The CrowdSearcher project [55, 58] focuses on *crowdsearching*, defined as an extension of crowdsourcing in which the human-assigned task is searching for information; the search can either be started ex novo, or it can be the refinement of a search result obtained from a search engine interaction.

Queries that can be crowdsourced may concern aspects such as getting fresh opinions about recent events, or suggestions about choices and dilemmas from a cluster of trusted experts or friends. The requestor knows that crowdsourced queries will not get an immediate answer and accepts to wait for a time which can vary from seconds to hours.

Crowdsourcing a search query brings about issues and problems which are different from crowdsourcing an operational task.

- Queries should be addressed to small groups of responders, and therefore *crowd selection* is quite critical; selection can be either manually performed by the requestor, or performed by an expertise matching system which extracts the query crowd from a repository of responders' profiles.
- Crowdsearching requires an *engagement platform*, i.e., a context where the requestor selects the crowd for a given query, and an *execution platform*, where each query is responded to. The two platforms can be different; e.g., users could be engaged by email and respond to a poll, managed by a system such as Doodle (www.doodle.com).

Application/Design dimension	Involvement			Ability			Task style			Control		Motivation			Time	
	Single	Small group	Open community	Perception & emotion	Judgement	Social behavior	Game	Explicit task	Implicit task	Central	Distributed	Fun	Volunteering	Reward	Critical	Noncritical
Crowdsourcing markets [57, 273]			✓	✓	✓			✓		✓				✓		✓
Algorithm optimization [345, 347]			✓		✓			✓		✓			✓	✓		✓
Image tagging/object location [161, 359]		✓	✓		✓		✓		✓	✓		✓				✓
User studies [...]	✓		✓													
Data collection																
Collective deliberation [102, 201]			✓		✓	✓										✓
Environmental monitoring [198, 237]																
Health care/disease control [235, 269]			✓		✓			✓	✓	✓			✓		✓	
Scientific data analysis					✓											
Emergency management [2, 134]			✓		✓	✓									✓	
Search [55, 357]																

Fig. 15.9 Applications of human computations and their design dimensions

(a) The crowd engagement mechanism of   (b) The crowd query answering interface of
CrowdSearcher   CrowdSearcher

Fig. 15.10   The CrowdSearcher project

- The *incentives* that move responders are not monetary, as responders are not performing a well-defined task with a clear ownership and remuneration. Responders
  may answer from a *genuine wish to be helpful*, as they desire to report a personal
  experience that may help someone; when the answer is shared within a community, a responder may increase *visibility* and *reputation*; when queries are framed
  within a question answering system where responders take part in a *game*, they
  respond for *fun*. Another factor that may influence responders is the *intensity of
  social activity of the asker*.

The CrowdSearcher system uses social networks, such as Facebook and LinkedIn,
as both engagement and execution platform. CrowdSearcher presents a uniform interface for defining search tasks, consisting of a collection of type search objects
which can be queried and manipulated with operations such as like, dislike, recommend, tag, score, order, group, add, delete, correct, and connect; a search object is
portable to the various social networks that play the role of execution platforms, and
can be implemented either with native controls (e.g., *like* actions on the requestor's
Facebook wall) or with embedded applications (deployed using the social network's
API). Figure 15.10(b) shows an example in which the requestor asks selected friends
their opinion about jobs found on the job market; Fig. 15.10(b) shows an example
of execution with Facebook native controls.

15.4.2 The CUbRIK Project

The CUbRIK project [130] is a large-scale integrating project partially funded by
the European Commission's 7th Framework ICT Programme. It aims at developing
a multimedia search framework to incorporate human and social computation and to
enrich the semantics of multimedia content and query processing with the support
of temporal and spatial entities.

The integration of machine and human tasks in CUbRIK exploits a distributed
system layered in four main tiers, as shown in Fig. 15.11.

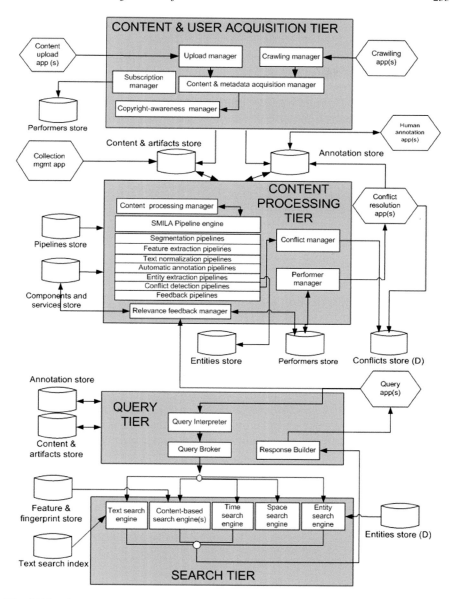

Fig. 15.11 The architecture of the CUbRIK system

- The *content and user acquisition tier* registers content and users into the system. The *Subscription Manager* handles two classes of users: searchers and performers. Searchers use CUbRIK applications to find information and provide feedback on query result quality. Performers execute tasks (via gaming or query & answer) to provide and validate semantic annotations and resolve conflicts produced by machine tasks.

- The *content processing tier* includes the *Content Processing Manager*, which listens to a queue of pending content processing requests and controls the corresponding tasks. Process control is implemented on top of the SMILA pipeline engine [114], which orchestrates Web service or local component calls for both general-purpose (e.g., video/audio segmentation) and domain-specific (e.g., face recognition) content processing logic. The output of a content processing task consists of derivative content (e.g., key frames, thumbnails, audio summaries), low-level features and high level annotations with their confidence values, and conflicts (low-confidence and contradictory annotations).

- The *Conflict Manager* is the core component for integrating HC; it manages low-confidence annotations (conflicts) and their assignment to machine applications and human performers. In the simplest case, conflicts can be assigned to an application, which manages their allocation to performers, possibly using data provided by the Performer Manager. This is the typical case for simple GWAP applications, where the interaction logic is user independent and only basic profile data, like the skill level of the gamer, are employed to decide which conflict to present. Alternatively, conflicts can be assigned to an application-performer pair: in this case, the association of the performer is managed by CUbRIK, and the application routes the conflict to the performer suggested by the platform. This is the case of more personalized applications, like query & answer, where a mix of history, profile, and trust data of the performer can be used to route the most appropriate questions. A conflict resolution application can also be an existing third-party application (e.g., a crowdsourcing application on top of a commercial platform). The Conflict Manager is also responsible for closing a conflict and storing the produced high-confidence annotations.

    The *Performer Manager* is responsible for keeping data about performers (profile, social network centrality measures, history of solved conflicts, throughput, quality of decision, etc.), which are used to optimize task allocation. Some pipelines are designed to receive feedback from the user on the results of a query. This feedback is routed to a *Relevance Feedback Manager* module that updates the level of trust of performers (human and automatic) in the component and performer store.

- The *query processing tier* consists of one or more *query applications* that contain the front end for issuing queries and viewing results. Queries are expressed according to a multimodal query language, serialized and submitted to a CUbRIK platform (through a Web services API); results are organized according to an application-dependent result schema, serialized, and returned as responses from CUbRIK to the application. Typically, results consist of references to matched content elements, match details for each content element (e.g., timestamp of match for continuous media), and annotation values for the collection of results (to enable faceted search).

Fig. 15.12 The interface of
the Sketchness GWAP for
clothes detection

- Finally, the *search tier* contains a collection of independent search engines. Each search engine can access the content and annotation store(s) to build/rebuild its indexes. Indexing is independent and asynchronous with respect to content processing and acquisition.

Two applications illustrate CUbRIK functionality. The first one, *History of Europe (HoE)*, builds upon the audio–visual content of a specialized digital library containing materials like TV/cinema news on history of Europe events, archives from institutional collections, interviews, and video materials. The HoE application enhances the audio–visual content by matching, with the help of the crowd, unclassified archive photos and videos with curated knowledge sources (e.g., repositories of portraits of historically relevant people and known events) in order to help researchers identify events and unknown people. The content enriching pipeline mixes human and social intelligence (e.g., digital humanities methodologies from researchers, and annotations and tags from groups of history experts and amateurs or from groups located in some specific European region) with machine learning techniques (text data analysis, automatic image classification, face detection, and face similarity computation).

The second application, *CUbRIK fashion trends (FT)*, focuses on searching fashion-related content. It indexes images crawled from both fashion-related content sharing sites and from user-generated content (UGC) repositories, crowdsources the accurate identification of clothes of various kinds by means of a game with a purpose (shown in Fig. 15.12), and automatically extracts useful style trends (e.g., color and shape evolution over time) from the human-annotated collection of objects extracted from the fashion photos.

15.5 Open Issues

Not only can machines fail or deliver low-quality output when addressing a computation task; humans also have limitations in their capacity to solve problems, and—unlike machines—can also make errors on purpose. The design of HC systems must take into consideration a number of issues that arise when humans are included in the loop [183].

- *Human Factors in Computing and Decision Making*: The human cognitive processes interplay with affective states and may exhibit systematic and predictable biases, which, if overlooked, may threaten the validity of human-assigned tasks. Human factors in decision making have been studied separately in neuroscience, behavioral economics, and cognitive psychology. Recently, the interdisciplinary field of neuroeconomics has addressed the integration of multiple techniques to examine decision making by considering cognitive and neural constraints, used in psychology and neuroscience, in the construction of mathematical decision models that are typical of economics [305]. Anchoring and sequential effects (the dependency of responses on prior information) have been studied (e.g., the "anchoring bias" studied by [333] and well known in the environmental decision-making literature), and mitigation measures have been proposed that could be applied to crowdsourcing and online judgment elicitation [258].
- *Quality of Human Output and Malicious Behavior*: A major issue in HC design is measuring and improving the quality of human contributions, which requires estimating the quality of the task performer and combining different (and potentially conflicting) results produced by different performers. The simplest technique for quality improvement is, when possible and sensible, to submit the same task to multiple performers and take the majority response as the true outcome, thus assimilating the quality measure to the probability that a response is correct [211]. Refined approaches rely on the estimation of the user's quality and combine the responses taking this estimation into account. As an extreme case, performers may intentionally cheat, producing incorrect results, which requires suitable techniques for detecting *spammers* [184], or they may simply be negligent, an attitude that can be detected by pre-task screening [110].
- *Market Design*: Some HC applications, e.g., task crowdsourcing for data validation, can be regarded as a work market, where work sellers and buyers meet [183]. As in any market, efficiency heavily depends on reciprocal trust between buyers and sellers, and can be hampered by opportunistic or fraudulent behaviors. Trust and reputation systems used in e-commerce and recommender systems [189] apply to HC too, but need to be adapted to the specificity of HC environments.
- *Ethical and Legal Aspects*: HC applications question the ethical and legal framework at the base of the human condition [374]. Many applications of HC (e.g., expert finding and task-to-worker matching) require processing the user's data, e.g., profiles and friendship links extracted from social networks.

These requirements may collide with the user's expectation about data protection and privacy. Furthermore, the disconnection of workers from employers implied by crowdsourcing is also changing the workplace relationships, with such potential consequences as worker's alienation, incapacity of judgment about the moral valence of tasks, and overmonitoring by the commissioner.

15.6 Exercises

15.1 Select a human computation application of your choice and classify its approach according to the design dimensions illustrated in Fig. 15.9.

15.2 Sketch a game concept applicable to the problem of classifying images into a binary category (e.g., indoor/outdoor).

References

1. T. Abdelzaher, Y. Anokwa, P. Boda, J. Burke, D. Estrin, L. Guibas, A. Kansal, S. Madden, J. Reich, Mobiscopes for human spaces. IEEE Pervasive Comput. **6**, 20–29 (2007)
2. F. Abel, C. Hauff, G.-J. Houben, R. Stronkman, K. Tao, Twitcident: fighting fire with information from social web streams, in *WWW (Companion Volume)* (2012), pp. 305–308
3. A. Abid, M. Tagliasacchi, Top-k query processing with parallel probing of data sources, in *SEBD*, ed. by G. Mecca, S. Greco (2011), p. 155
4. S. Abney, M. Collins, A. Singhal, Answer extraction, in *Proceedings of the Sixth Conference on Applied Natural Language Processing. ANLC'00* (Association for Computational Linguistics, Stroudsburg, 2000), pp. 296–301
5. B. Adida, M. Birbeck, S. McCarron, I. Herman, RDFa Core 1.1—syntax and processing rules for embedding RDF through attributes, Technical report, W3C Recommendation, 2010. http://www.w3.org/TR/rdfa-core/
6. B. Adida, M. Birbeck, S. Pemberton, HTML+RDFa 1.1—support for RDFa in HTML4 and HTML5, Technical report, W3C Working Draft, 2012. http://www.w3.org/TR/rdfa-in-html/
7. B. Aditya, G. Bhalotia, S. Chakrabarti, A. Hulgeri, C. Nakhe, P. Parag, S. Sudarshan, BANKS: browsing and keyword searching in relational databases, in *VLDB 2002, Proceedings of 28th International Conference on Very Large Data Bases*, 20–23 August, 2002, Hong Kong, China (2002), pp. 1083–1086
8. Adobe Inc., Adobe Premiere, http://www.adobe.com/products/premiere. Accessed Sept 2012
9. G. Aggarwal, J. Feldman, S. Muthukrishnan, M. Pál, Sponsored search auctions with Markovian users, in *WINE* (2008), pp. 621–628
10. G. Aggarwal, S. Muthukrishnan, D. Pál, M. Pál, General auction mechanism for search advertising, in *WWW* (2009), pp. 241–250
11. L. Agnihotri, N. Dimitrova, J.R. Kender, J. Zimmerman, Study on requirement specifications for personalized multimedia summarization, in *Proceedings of the 2003 IEEE International Conference on Multimedia and Expo*, ICME 2003, 6–9 July 2003, Baltimore, MD, USA (2003), pp. 757–760
12. S. Agrawal, S. Chaudhuri, G. Das, Dbxplorer: a system for keyword-based search over relational databases, in *ICDE*, ed. by R. Agrawal, K.R. Dittrich (IEEE Comput. Soc., Los Alamitos, 2002), pp. 5–16
13. A. Ahmad, L. Dey, A method to compute distance between two categorical values of same attribute in unsupervised learning for categorical data set. Pattern Recognit. Lett. **28**(1), 110–118 (2007)
14. R. Albert, H. Jeong, A.L. Barabasi, The diameter of the world wide web. Nature **401**, 130–131 (1999)
15. R.B. Allen, Retrieval from facet spaces. Electron. Publ. **8**, 247–257 (1995)

S. Ceri et al., *Web Information Retrieval*, Data-Centric Systems and Applications, DOI 10.1007/978-3-642-39314-3, © Springer-Verlag Berlin Heidelberg 2013

16. J. Allsopp, Link-based microformats: rel-license, rel-tag, rel-nofollow, and votelinks, in *Microformats: Empowering Your Markup for Web 2.0* (Apress, New York, 2007), pp. 51–74

17. G. Alonso, F. Casati, H. Kuno, V. Machiraju, *Web Services—Concepts, Architectures and Applications*, 1st edn. (Springer, Berlin, 2003)

18. O. Alonso, M. Gertz, R.A. Baeza-Yates, On the value of temporal information in information retrieval. SIGIR Forum **41**(2), 35–41 (2007)

19. M. Ankerst, M.M. Breunig, H.P. Kriegel, J. Sander, Optics: ordering points to identify the clustering structure, in *ACM SIGMOD International Conference on Management of Data* (1999)

20. M.A. Anusuya, S.K. Katti, Speech recognition by machine, a review. Int. J. Comput. Sci. Inform. Secur. **6**(3), 181–205 (2009)

21. A. Arasu, J. Cho, H. Garcia-Molina, A. Paepcke, S. Raghavan, Searching the web. ACM Trans. Internet Technol. **1**(1), 2–43 (2001)

22. A. Aula, D.M. Russell, Complex and exploratory web search, in *Information Seeking Support Systems* (2008)

23. O. Aulov, M. Halem, Assimilation of real-time satellite and human sensor networks for modeling natural disasters, in *AGU Fall Meeting Abstracts* (2011)

24. C. Badue, B. Ribeiro-Neto, R. Baeza-Yates, N. Ziviani, Distributed query processing using partitioned inverted files, in *Proceedings of the Eighth International Symposium on String Processing and Information Retrieval*, SPIRE 2001, Nov (2001), pp. 10–20

25. R.A. Baeza-Yates, B. Ribeiro-Neto, *Modern Information Retrieval* (Addison-Wesley, Boston, 1999)

26. R.A. Baeza-Yates, B.A. Ribeiro-Neto, *Modern Information Retrieval—the Concepts and Technology Behind Search*, 2nd edn. (Pearson Education, Harlow, 2011)

27. R. Baeza-Yates, F. Saint-Jean, A three level search engine index based in query log distribution, in *String Processing and Information Retrieval*, ed. by M. Nascimento, E. Moura, A. Oliveira. Lecture Notes in Computer Science, vol. 2857 (Springer, Berlin, 2003), pp. 56–65

28. R. Baeza-Yates, C. Castillo, M. Marin, A. Rodriguez, Crawling a country: better strategies than breadth-first for web page ordering, in *Special Interest Tracks and Posters of the 14th International Conference on World Wide Web*. WWW'05 (ACM, New York, 2005), pp. 864–872

29. R.A. Baeza-Yates, C. Castillo, F. Junqueira, V. Plachouras, F. Silvestri, Challenges on distributed web retrieval, in *Proceedings of the 23rd International Conference on Data Engineering*, ICDE 2007, The Marmara Hotel, Istanbul, Turkey, ed. by R. Chirkova, A. Dogac, M.T. Özsu, T.K. Sellis (IEEE Press, New York, 2007), pp. 6–20

30. R. Baeza-Yates, A. Gionis, F. Junqueira, V. Murdock, V. Plachouras, F. Silvestri, The impact of caching on search engines, in *Proceedings of the 30th Annual International ACM SIGIR Conference on Research and Development in Information Retrieval*. SIGIR'07 (ACM, New York, 2007), pp. 183–190

31. R. Baeza-Yates, A. Gionis, F.P. Junqueira, V. Murdock, V. Plachouras, F. Silvestri, Design trade-offs for search engine caching. ACM Trans. Web **2**(4), 20 (2008)

32. R. Baeza-Yates, A. Gionis, F. Junqueira, V. Plachouras, L. Telloli, On the feasibility of multisite web search engines, in *Proceedings of the 18th ACM Conference on Information and Knowledge Management*. CIKM'09 (ACM, New York, 2009), pp. 425–434

33. A. Baldi, R. Murace, E. Dragonetti, M. Manganaro, O. Guerra, S. Bizzi, L. Galli, Definition of an automated content-based image retrieval (CBIR) system for the comparison of dermoscopic images of pigmented skin lesions. Biomed. Eng. Online **8**, 18 (2009)

34. N. Bansal, B. Srivastava, On using crowd for measuring traffic at aggregate level for emerging countries, in *Proceedings of the 8th International Workshop on Information Integration on the Web: in Conjunction with WWW 2011* (2011), pp. 5:1–5:6

35. J. Barker, C. Hennesy, Meta-search engines (2012), http://www.lib.berkeley.edu/TeachingLib/Guides/Internet/MetaSearch.html

36. L.A. Barroso, J. Dean, U. Hölzle, Web search for a planet: the Google cluster architecture. IEEE MICRO **23**(2), 22–28 (2003)
37. M.J. Bates, Information search tactics. J. Am. Soc. Inf. Sci. **30**(4), 205–214 (1979)
38. M.J. Bates, The design of browsing and berrypicking techniques for the online search interface. Online Inf. Rev. **13**(5), 407–424 (1989)
39. L. Becchetti, C. Castillo, The distribution of pagerank follows a power-law only for particular values of the damping factor, in *Proceedings of the 15th International Conference on World Wide Web*. WWW'06 (ACM, New York, 2006), pp. 941–942
40. D. Beeferman, A. Berger, J. Lafferty, Statistical models for text segmentation. Mach. Learn. **34**, 177–210 (1999) doi:10.1023/A:1007506220214
41. S.M. Beitzel, E.C. Jensen, D.A. Grossman, Retrieving OCR text: a survey of current approaches, in *Symposium on Document Image Understanding Technologies (SDUIT)* (2003)
42. T. Belluf, L. Xavier, R. Giglio, Case study on the business value impact of personalized recommendations on a large online retailer, in *Proceedings of the Sixth ACM Conference on Recommender Systems*. RecSys'12 (ACM, New York, 2012), pp. 277–280
43. J.P. Benway, Banner blindness: the irony of attention grabbing on the world wide web. Proc. Hum. Factors Ergon. Soc. Annu. Meet. **42**(5), 463–467 (1998)
44. M.K. Bergman, White paper: the deep web. Surfacing hidden value. J. Electron. Publ. **7**(1) (2001). doi:10.3998/3336451.0007.104
45. T. Berners-Lee, Linked data, World wide web design issues (2010), http://www.w3.org/DesignIssues/LinkedData.html
46. K. Bharat, M.R. Henzinger, Improved algorithms for topic distillation in a hyperlinked environment, in *Proceedings of the 21st Annual International ACM SIGIR Conference on Research and Development in Information Retrieval*. SIGIR'98 (ACM, New York, 1998), pp. 104–111
47. C. Bizer, The emerging web of linked data. IEEE Intell. Syst. **24**(5), 87–92 (2009)
48. C. Bizer, T. Heath, T. Berners-Lee, Linked data—the story so far. Int. J. Semantic Web Inf. Syst. **5**(3), 1–22 (2009)
49. C. Bizer, J. Lehmann, G. Kobilarov, S. Auer, C. Becker, R. Cyganiak, S. Hellmann, DBpedia—a crystallization point for the web of data. J. Web Semant. **7**(3), 154–165 (2009)
50. R. Blanco, E. Bortnikov, F. Junqueira, R. Lempel, L. Telloli, H. Zaragoza, Caching search engine results over incremental indices, in *Proceedings of the 33rd International ACM SIGIR Conference on Research and Development in Information Retrieval*. SIGIR'10 (ACM, New York, 2010), pp. 82–89
51. A. Borodin, G.O. Roberts, J.S. Rosenthal, P. Tsaparas, Link analysis ranking: algorithms, theory, and experiments. ACM Trans. Internet Technol. **5**(1), 231–297 (2005)
52. A. Bozzon, M. Brambilla, P. Fraternali, F.S. Nucci, S. Debald, E. Moore, W. Nejdl, M. Plu, P. Aichroth, O. Pihlajamaa, C. Laurier, S. Zagorac, G. Backfried, D. Weinland, V. Croce, Pharos: an audiovisual search platform, in *Proceedings of the 32nd Annual International ACM SIGIR Conference on Research and Development in Information Retrieval*, SIGIR 2009, July 19–23, 2009, Boston, MA, USA (2009), p. 841
53. A. Bozzon, M. Brambilla, S. Ceri, P. Fraternali, Liquid query: multi-domain exploratory search on the web, in *WWW*, ed. by M. Rappa, P. Jones, J. Freire, S. Chakrabarti (ACM, New York, 2010), pp. 161–170
54. A. Bozzon, M. Brambilla, S. Ceri, S. Quarteroni, A framework for integrating, exploring, and searching location-based web data. IEEE Internet Comput. **15**(6), 24–31 (2011)
55. A. Bozzon, M. Brambilla, S. Ceri, Answering search queries with CrowdSearcher, in *WWW*, ed. by A. Mille, F.L. Gandon, J. Misselis, M. Rabinovich, S. Staab (ACM, New York, 2012), pp. 1009–1018
56. A. Bozzon, M. Brambilla, P. Fraternali, M. Tagliasacchi, Diversification for multi-domain result sets, in *Web Engineering*, ed. by M. Brambilla, T. Tokuda, R. Tolksdorf, Lecture Notes in Computer Science, vol. 7387 (Springer, Berlin, 2012), pp. 137–152
57. A. Bozzon, I. Catallo, E. Ciceri, P. Fraternali, D. Martinenghi, M. Tagliasacchi, A framework for crowdsourced multimedia processing and querying, in *CrowdSearch*, ed. by R.A. Baeza-

Yates, S. Ceri, P. Fraternali, F. Giunchiglia. CEUR Workshop Proceedings, vol. 842 (2012), pp. 42–47. CEUR-WS.org

58. A. Bozzon, M. Brambilla, S. Ceri, A. Mauri, Reactive crowdsourcing, in *WWW* (2013), pp. 153–164

59. D. Braga, S. Ceri, F. Daniel, D. Martinenghi, Optimization of multi-domain queries on the web. Proc. VLDB Endow. **1**(1), 562–573 (2008)

60. M. Brambilla, M. Zanoni, Clustering and labeling of multi-dimensional mixed structured data, in *Search Computing*, ed. by S. Ceri, M. Brambilla. Lecture Notes in Computer Science, vol. 7538 (Springer, Berlin, 2012), pp. 111–126

61. M. Brambilla, P. Fraternali, C. Vaca, BPMN and design patterns for engineering social BPM solutions, in *Proceedings of the 4th International Workshop on BPM and Social Software (BPMS 2011)* (2011)

62. M. Brambilla, S. Ceppi, N. Gatti, E.H. Gerding, A revenue sharing mechanism for federated search and advertising, in *Proceedings of the 21st International Conference Companion on World Wide Web*. WWW'12 Companion (ACM, New York, 2012), pp. 465–466

63. D. Brickley, R.V. Guha, RDF vocabulary description language 1.0: RDF schema, Technical report, W3C Recommendation (2004), http://www.w3.org/TR/rdf-schema/

64. S. Brin, L. Page, The anatomy of a large-scale hypertextual web search engine. Comput. Netw. ISDN Syst. **30**(1–7), 107–117 (1998)

65. A. Broder, A taxonomy of web search. SIGIR Forum **36**(2), 3–10 (2002)

66. A.Z. Broder, S.C. Glassman, M.S. Manasse, G. Zweig, Syntactic clustering of the web. Comput. Netw. ISDN Syst. **29**(8–13), 1157–1166 (1997)

67. A. Broder, R. Kumar, F. Maghoul, P. Raghavan, S. Rajagopalan, R. Stata, A. Tomkins, J. Wiener, Graph structure in the web, in *Proceedings of the 9th International World Wide Web Conference on Computer Networks: the International Journal of Computer and Telecommunications Networking* (North-Holland, Amsterdam, 2000), pp. 309–320

68. C. Buckley, Implementation of the SMART information retrieval system, Technical report, Cornell University, Ithaca, NY, USA, 1985

69. W.L. Buntine, J. Löfström, J. Perkiö, S. Perttu, V. Poroshin, T. Silander, H. Tirri, A.J. Tuominen, V.H. Tuulos, A scalable topic-based open source search engine, in *2004 IEEE/WIC/ACM International Conference on Web Intelligence (WI 2004)*, 20–24 September 2004, Beijing, China, 2004, pp. 228–234

70. G. Buscher, E. Cutrell, M.R. Morris, What do you see when you're surfing?: using eye tracking to predict salient regions of web pages, in *Proceedings of the 27th International Conference on Human Factors in Computing Systems*. CHI'09 (ACM, New York, 2009), pp. 21–30

71. O.J. Cacho, D. Spring, S. Hester, R.M. Nally, Allocating surveillance effort in the management of invasive species: a spatially-explicit model. Environ. Model. Softw. **25**(4), 444–454 (2011)

72. D. Calvanese, G.D. Giacomo, D. Lembo, M. Lenzerini, R. Rosati, Tractable reasoning and efficient query answering in description logics: the *dl-lite* family. J. Autom. Reason. **39**(3), 385–429 (2007)

73. B.B. Cambazoglu, F.P. Junqueira, V. Plachouras, S. Banachowski, B. Cui, S. Lim, B. Bridge, A refreshing perspective of search engine caching, in *Proceedings of the 19th International Conference on World Wide Web*. WWW'10 (ACM, New York, 2010), pp. 181–190

74. A.T. Campbell, S.B. Eisenman, N.D. Lane, E. Miluzzo, R.A. Peterson, H. Lu, X. Zheng, M. Musolesi, K. Fodor, G.-S. Ahn, The rise of people-centric sensing. IEEE Internet Comput. **12**(4), 12–21 (2008)

75. J. Carbonell, J. Goldstein, The use of MMR, diversity-based reranking for reordering documents and producing summaries, in *SIGIR'98: Proceedings of the 21st Annual International ACM SIGIR Conference on Research and Development in Information Retrieval* (ACM, New York, 1998), pp. 335–336

76. A. Carlson, C. Cumby, J. Rosen, D. Roth, The SNoW learning architecture, Technical report, Technical report UIUCDCS, 1999

77. X. Carreras, L. Màrquez, Introduction to the CoNLL-2005 shared task: semantic role labeling, in *Proceedings of the Ninth Conference on Computational Natural Language Learning*, (Association for Computational Linguistics, Stroudsburg, 2005), pp. 152–164

78. C. Castillo, M. Marin, A. Rodriguez, R. Baeza-Yates, Scheduling algorithms for web crawling, in *Proceedings of the WebMedia & LA-Web 2004 Joint Conference 10th Brazilian Symposium on Multimedia and the Web 2nd Latin American Web Congress*. LA-WEBMEDIA'04 (IEEE Comput. Soc., Washington, 2004), pp. 10–17

79. R. Cavallo, Optimal decision-making with minimal waste: strategyproof redistribution of VCG payments, in *AAMAS*, (ACM, New York, 2001), pp. 882–889

80. W.B. Cavnar, J.M. Trenkle, N-gram-based text categorization. Ann Arbor MI 48113(2), 161–175 (1994), http://citeseerx.ist.psu.edu/viewdoc/summary?doi=10.1.1.53.9367

81. I. Celino, E.D. Valle, D. Cerizza, A. Turati, Squiggle: an experience in model-driven development of real-world semantic search engines, in *ICWE*, ed. by L. Baresi, P. Fraternali, G.-J. Houben. Lecture Notes in Computer Science, vol. 4607 (Springer, Berlin, 2007), pp. 485–490

82. S. Ceppi, N. Gatti, E.H. Gerding, Mechanism design for federated sponsored search auctions, in *AAAI* (2011)

83. S. Ceri, M. Brambilla (eds.), *Search Computing—Trends and Developments [Outcome of the Second SeCO Workshop on Search Computing, Como/Milan, Italy, May 25–31, 2010]*. Lecture Notes in Computer Science, vol. 6585 (Springer, Berlin, 2011)

84. S. Ceri, D. Braga, M. Brambilla, A. Campi, E. Della Valle, P. Fraternali, D. Martinenghi, M. Tagliasacchi, *Method for extracting, merging and ranking search engine results*, US Patent US 8,180,768 B2, May, 2012. http://www.google.com/patents/US8180768

85. S. Chakrabarti, *Mining the Web: Discovering Knowledge from Hypertext Data* (Morgan Kauffman, San Mateo, 2002)

86. K.T. Chan, I. King, M.-C. Yuen, Mathematical modeling of social games, in *International Conference on Computational Science and Engineering*, CSE'09, vol. 4 (2009), pp. 1205–1210

87. E. Charniak, Statistical techniques for natural language parsing. AI Mag. **18**, 33–44 (1997)

88. P.P. Chen, The entity-relationship model—toward a unified view of data. ACM Trans. Database Syst. **1**(1), 9–36 (1976)

89. Y. Chen, A. Ghosh, R. McAfee, D. Pennock, Sharing online advertising revenue with consumers, in *WINE* (2008), pp. 556–565

90. J. Cho, H. Garcia-Molina, L. Page, Efficient crawling through URL ordering, in *Proceedings of the Seventh International Conference on World Wide Web 7*, WWW7 (Elsevier, Amsterdam, 1998), pp. 161–172

91. C.W. Cleverdon, *Report on the Testing and Analysis of an Investigation into the Comparative Efficiency of Indexing Systems* (1962)

92. M. Collins, Head-driven statistical models for natural language parsing, Ph.D. thesis, University of Pennsylvania, 1999

93. M. Collins, N. Duffy, New ranking algorithms for parsing and tagging: kernels over discrete structures, and the voted perceptron, in *ACL* (2002)

94. K. Collins-Thompson, J. Callan, A language modeling approach to predicting reading difficulty, in *Proceedings of HLT/NAACL*, vol. 4 (2004)

95. S. Cooper, F. Khatib, A. Treuille, J. Barbero, J. Lee, M. Beenen, A. Leaver-Fay, D. Baker, Z. Popovic, Predicting protein structures with a multiplayer online game. Nature **466**(7307), 756–760 (2010)

96. J. Cope, N. Craswell, D. Hawking, Automated discovery of search interfaces on the web, in *ADC*, ed. by K.-D. Schewe, X. Zhou. CRPIT, vol. 17 (Australian Computer Society, Oarlinghurst, 2003), pp. 181–189

97. C. Cotsaces, N. Nikolaidis, I. Pitas, Video shot detection and condensed representation. A review. IEEE Signal Process. Mag. **23**(2), 28–37 (2006). doi:10.1109/MSP.2006.1621446

98. W.B. Croft, D.J. Harper, Using probabilistic models of document retrieval without relevance information, in *Document Retrieval Systems*, ed. by P. Willett (Taylor Graham Publishing, London, 1988), pp. 161–171

99. A. Culotta, J. Sorensen, Dependency tree kernels for relation extraction, in *Proceedings of the 42nd Annual Meeting on Association for Computational Linguistics* (Association for Computational Linguistics, Stroudsburg, 2004), p. 423

100. D. Cutting, J. Pedersen, Optimization for dynamic inverted index maintenance, in *Proceedings of the 13th Annual International ACM SIGIR Conference on Research and Development in Information Retrieval*, SIGIR'90 (ACM, New York, 1990), pp. 405–411

101. DCMI, Dublin core metadata initiative, http://dublincore.org. Accessed Sept 2012

102. F. De Cindio, E. Krzatala-Jaworska, L. Sonnante, Problems&Proposals: a tool for collecting citizens intelligence, in *Proc. of the Second International Workshop on Collective Intelligence at ACM CSCW Conference*, ed. by ACM, Seattle, Washington, February (2012)

103. E. Della Valle, I. Celino, D. Dell'Aglio, The experience of realizing a semantic web urban computing application. Trans. GIS **14**(2), 163–181 (2010)

104. B. Delvin, J. Wilkinson, The material exchange format, in *File Interchange Handbook*, ed. by B. Gilmer (Elsevier/Focal Press, Amsterdam/Burlington, 2004)

105. E. Demidova, P. Fankhauser, X. Zhou, W. Nejdl, DivQ: diversification for keyword search over structured databases, in *SIGIR'10: Proceeding of the 33rd International ACM SIGIR Conference on Research and Development in Information Retrieval* (ACM, New York, 2010), pp. 331–338

106. N.R. Devanur, S.M. Kakade, The price of truthfulness for pay-per-click auctions, in *ACM EC* (2009), pp. 99–106

107. M.B. Dias, D. Locher, M. Li, W. El-Deredy, P.J.G. Lisboa, The value of personalised recommender systems to e-business: a case study, in *Proceedings of the 2008 ACM Conference on Recommender Systems*. RecSys'08 (ACM, New York, 2008), pp. 291–294

108. L.R. Dice, Measures of the amount of ecologic association between species. Ecology **26**(3), 297–302 (1945)

109. Z. Dou, S. Hu, K. Chen, R. Song, J.-R. Wen, Multi-dimensional search result diversification, in *Proceedings of the Fourth ACM International Conference on Web Search and Data Mining*, WSDM'11 (ACM, New York, 2011), pp. 475–484

110. J.S. Downs, M.B. Holbrook, S. Sheng, L.F. Cranor, Are your participants gaming the system? Screening mechanical turk workers, in *Proceedings of the 28th International Conference on Human Factors in Computing Systems* (2010), pp. 2399–2402

111. M. Drosou, E. Pitoura, Search result diversification. SIGMOD Rec. **39**(1), 41–47 (2010)

112. R. Duda, P. Hart, *Pattern Classification and Scene Analysis* (Wiley, New York, 1973)

113. P. Dutta, P.M. Aoki, N. Kumar, A. Mainwaring, C. Myers, W. Willett, A. Woodruff, Common Sense: Participatory urban sensing using a network of handheld air quality monitors, in *Proceedings of the 7th ACM Conference on Embedded Networked Sensor Systems* (2009), pp. 349–350

114. Eclipse Foundation, SMILA—SeMantic Information Logistics Architecture, http://www.eclipse.org/smila. Accessed Sept 2012

115. B. Edelman, M. Ostrovsky, M. Schwarz, Internet advertising and the generalized second price auction: selling billions of dollars worth of keywords. NBER Working paper 11765 (2005)

116. B. Edelman, M. Ostrovsky, M. Schwarz, Internet advertising and the generalized second-price auction: selling billions of dollars worth of keywords. Am. Econ. Rev. **97**(1), 242–259 (2007)

117. S. Elbassuoni, M. Ramanath, R. Schenkel, M. Sydow, G. Weikum, Language-model-based ranking for queries on RDF-graphs, in *CIKM*, ed. by D.W.-L. Cheung, I.-Y. Song, W.W. Chu, X. Hu, J.J. Lin (ACM, New York, 2009), pp. 977–986

118. D.W. Embley, Y.S. Jiang, Y.-K. Ng, Record-boundary discovery in web documents, in *SIGMOD Conference*, ed. by A. Delis, C. Faloutsos, S. Ghandeharizadeh (ACM, New York, 1999), pp. 467–478

119. J. English, M. Hearst, R. Sinha, K. Swearingen, K.-P. Yee, Hierarchical faceted metadata in site search interfaces, in *CHI'02 Extended Abstracts on Human Factors in Computing Systems*. CHI EA'02 (ACM, New York, 2002), pp. 628–639

120. F. Esposito, N. Fanizzi, C. d'Amato, *Partitional Conceptual Clustering of Web Resources Annotated with Ontology Languages* (Springer, Berlin, 2009)

121. A. Esuli, F. Sebastiani, Determining term subjectivity and term orientation for opinion mining, in *Proceedings the 11th Meeting of the European Chapter of the Association for Computational Linguistics (EACL-2006)* (2006), pp. 193–200

122. Eyealike, Inc., Eyealike Platform for Online Shopping, http://www.eyealike.com/home. Accessed Sept 2012

123. R. Fagin, A. Lotem, M. Naor, Optimal aggregation algorithms for middleware, in *PODS*, ed. by P. Buneman (ACM, New York, 2001)

124. T. Fagni, R. Perego, F. Silvestri, S. Orlando, Boosting the performance of web search engines: caching and prefetching query results by exploiting historical usage data. ACM Trans. Inf. Syst. **24**(1), 51–78 (2006)

125. B. Fazzinga, T. Lukasiewicz, Semantic search on the web. Semant. Web **1**(1), 89–96 (2010)

126. B. Fazzinga, G. Gianforme, G. Gottlob, T. Lukasiewicz, Semantic web search based on ontological conjunctive queries. J. Web Semant. **9**(4), 453–473 (2011)

127. M. Fernández, I. Cantador, V. Lopez, D. Vallet, P. Castells, E. Motta, Semantically enhanced information retrieval: an ontology-based approach. J. Web Semant. **9**(4), 434–452 (2011)

128. D.H. Fisher, Knowledge acquisition via incremental conceptual clustering. Mach. Learn. **2**, 139–172 (1987)

129. P. Fraternali, M. Brambilla, A. Bozzon, Model-driven design of audiovisual indexing processes for search-based applications, in *Seventh International Workshop on Content-Based Multimedia Indexing*, CBMI'09, 3–5 June, 2009, Chania, Crete (2009), pp. 120–125

130. P. Fraternali, M. Tagliasacchi, D. Martinenghi, A. Bozzon, I. Catallo, E. Ciceri, F.S. Nucci, V. Croce, I.S. Altingövde, W. Siberski, F. Giunchiglia, W. Nejdl, M. Larson, E. Izquierdo, P. Daras, O. Chrons, R. Traphöner, B. Decker, J. Lomas, P. Aichroth, J. Novak, G. Sillaume, F. Sánchez-Figueroa, C. Salas-Parra, The CUBRIK project: human-enhanced time-aware multimedia search, in *WWW (Companion Volume)*, ed. by A. Mille, F.L. Gandon, J. Misselis, M. Rabinovich, S. Staab (ACM, New York, 2012), pp. 259–262

131. N. Fuhr, Probabilistic models in information retrieval. Comput. J. **35**(3), 243–255 (1992)

132. W.K. Gad, M.S. Kamel, Incremental clustering algorithm based on phrase-semantic similarity histogram, in *Proceedings of the Ninth International Conference on Machine Learning and Cybernetics*, Qingdao (2010)

133. G. Gan, C. Ma, J. Wu, *Data Clustering: Theory, Algorithms, and Applications* (SIAM, Philadelphia, 2007)

134. H. Gao, G. Barbier, R. Goolsby, Harnessing the crowdsourcing power of social media for disaster relief. IEEE Intell. Syst. **26**(3), 10–14 (2011)

135. N. Gatti, A. Lazaric, F. Trovo, A truthful learning mechanism for contextual multi-slot sponsored search auctions with externalities, in *ACM EC* (2012)

136. N. Gatti, A. Lazaric, F. Trovo, A truthful learning mechanism for multi-slot sponsored search auctions with externalities, in *AAMAS* (2012)

137. J. Geurts, J. van Ossenbruggen, L. Hardman, Requirements for practical multimedia annotation, in *Proc. of the Int. Workshop on Multimedia and the Semantic Web*, Heraklion, Crete (2005), pp. 4–11

138. A. Ghosh, P. McAfee, K. Papineni, S. Vassilvitskii, Bidding for representative allocations for display advertising, in *WINE* (2009), pp. 208–219

139. F. Giunchiglia, U. Kharkevich, I. Zaihrayeu, Concept search, in *ESWC*, ed. by L. Aroyo, P. Traverso, F. Ciravegna, P. Cimiano, T. Heath, E. Hyvönen, R. Mizoguchi, E. Oren, M. Sabou, E.P.B. Simperl Lecture Notes in Computer Science, vol. 5554 (Springer, Berlin, 2009), pp. 429–444

140. A. Goel, K. Munagala, Hybrid keyword search auctions, in *WWW* (2009), pp. 221–230

141. S. Gollapudi, A. Sharma, An axiomatic approach for result diversification, in *WWW'09: Proceedings of the 18th International Conference on World Wide Web* (ACM, New York, 2009), pp. 381–390

142. D.C. Gondek, A. Lally, A. Kalyanpur, J.W. Murdock, P.A. Duboue, L. Zhang, Y. Pan, Z.M. Qiu, C. Welty, A framework for merging and ranking of answers in DeepQA. IBM J. Res. Dev. **56**(3), 399–410 (2012)

143. T.F. Gonzalez, Clustering to minimize the maximum intercluster distance. Theor. Comput. Sci. **38**, 293–306 (1985)

144. J. Gonzalo, F. Verdejo, I. Chugur, J.M. Cigarrán, Indexing with WordNet synsets can improve text retrieval. In *Proceedings of the COLING/ACL'98 Workshop on Usage of WordNet for NLP*, Montreal (1998)

145. Google Inc., Google election video search (2009). http://googleblog.blogspot.com/2008/07/in-their-own-words-political-videos.html

146. Google Inc., Google Picasa, http://picasa.google.com. Accessed Sept 2012

147. J. Green, J.J. Laffont, Characterization of satisfactory mechanisms for the revelation of preferences for public goods. Econometrica **45**, 427–438 (1977)

148. S. Grimes, Unstructured data and the 80 percent rule. Carabridge Bridgepoints (2008)

149. D.A. Grossman, O. Frieder, *Information Retrieval: Algorithms and Heuristics*, vol. 15 (Kluwer Academic, Norwell, 2004)

150. R.V. Guha, R. McCool, E. Miller, Semantic search, in *WWW* (2003), pp. 700–709

151. M. Guo, V. Conitzer, Undominated VCG redistribution mechanisms, in *AAMAS* (2008), pp. 1039–1046

152. M. Hamilton, F. Salim, E. Cheng, S. Choy, Transafe: a crowdsourced mobile platform for crime and safety perception management. Comput. Soc. **41**(2), 32–37 (2011)

153. U. Hanani, B. Shapira, P. Shoval, Information filtering: overview of issues, research and systems. User Model. User-Adapt. Interact. **11**(3), 203–259 (2001)

154. A. Hanbury, A survey of methods for image annotation. J. Vis. Lang. Comput. **19**(5), 617–627 (2008)

155. A. Harth, VisiNav: a system for visual search and navigation on web data. J. Web Semant. **8**(4), 348–354 (2010)

156. T. Hastie, R. Tibshirani, J.H. Friedman, The elements of statistical learning, in *Boosting and Additive Trees*, 2nd edn. (Springer, New York, 2009), pp. 337–384

157. K.-i. Hatsuda, M. Yonezawa, T. Nonaka, T. Hase, Analysis of psychological movement of lines of sight when using Internet search engines, in *IEEE 13th International Symposium on Consumer Electronics, 2009*, ISCE'09 (2009), pp. 586–587

158. B. He, M. Patel, Z. Zhang, K.C.-C. Chang, Accessing the deep web. Commun. ACM **50**(5), 94–101 (2007)

159. H.S. Heaps, *Information Retrieval: Computational and Theoretical Aspects* (Academic Press, San Diego, 1978)

160. M.A. Hearst, *Search User Interfaces*, 1st edn. (Cambridge University Press, New York, 2009)

161. J. Heer, M. Bostock, Crowdsourcing graphical perception: using mechanical turk to assess visualization design, in *Proceedings of the 28th International Conference on Human Factors in Computing Systems*, New York, NY (2010), pp. 203–212

162. W. Heeren, L. van der Werff, R. Ordelman, A. van Hessen, F. de Jong, Radio Oranje: searching the queen's speech(es), in *SIGIR*, ed. by W. Kraaij, A.P. de Vries, C.L.A. Clarke, N. Fuhr, N. Kando (ACM, New York, 2007), p. 903

163. J. Heflin, J.A. Hendler, S. Luke, SHOE: a blueprint for the semantic web, in *Spinning the Semantic Web*, ed. by D. Fensel, J.A. Hendler, H. Lieberman, W. Wahlster (MIT Press, Cambridge, 2003), pp. 29–63

164. M. Hepp, GoodRelations: an ontology for describing products and services offers on the web, in *EKAW*, ed. by A. Gangemi, J. Euzenat. Lecture Notes in Computer Science, vol. 5268 (Springer, Berlin, 2008), pp. 329–346

165. A. Heydon, M. Najork, Mercator: a scalable, extensible web crawler. World Wide Web J. **2**(4), 219–229 (1999)

166. J. Hoffart, F.M. Suchanek, K. Berberich, E. Lewis-Kelham, G. de Melo, G. Weikum, YAGO2: exploring and querying world knowledge in time, space, context, and many languages, in *WWW (Companion Volume)*, ed. by S. Srinivasan, K. Ramamritham, A. Kumar, M.P. Ravindra, E. Bertino, R. Kumar (ACM, New York, 2011), pp. 229–232

167. D.W. Hosmer, S. Lemeshow, *Applied Logistic Regression*, vol. 354 (Wiley-Interscience, New York, 2000)

168. G. Hotchkiss, S. Alston, G. Edwards, Google eye tracking report. Enquiro Eyetools and DidIt (July), 1–106 (2005). http://pages.enquiro.com/whitepaper-enquiro-eye-tracking-report-I-google.html

169. G. Hotchkiss, T. Sherman, R. Tobin, C. Bates, K. Brown, Search engine results: 2010, Technical report, Enquiro Research, 2007

170. J. Howe, The rise of crowdsourcing. Wired **14**(6) (2006)

171. V. Hristidis, Y. Papakonstantinou, DISCOVER: keyword search in relational databases, in *VLDB 2002, Proceedings of 28th International Conference on Very Large Data Bases*, 20–23 Aug 2002, Hong Kong, China (2002), pp. 670–681

172. C.-C. Hsu, C.-L. Chen, Y.-W. Su, Hierarchical clustering of mixed data based on distance hierarchy. Inf. Sci. **177**, 4474–4492 (2007)

173. X. Hu, M. Stalnacke, T.B. Minde, R. Carlsson, S. Larsson, A mobile game to collect and improve position of images, in *Proceedings of the 3rd International Conference on Next Generation Mobile Applications, Services and Technologies* (2009)

174. Z. Huang, A fast clustering algorithm to cluster very large categorical data sets in data mining, in *Proceedings of the SIGMOD Workshop on Research Issues on Data Mining and Knowledge Discovery*, Dept. of Computer Science, The University of British Columbia, Canada (1997)

175. Z. Huang, Extensions to the k-means algorithm for clustering large data sets with categorical values. Data Min. Knowl. Discov. **2**, 283–304 (1998)

176. X. Huang, A. Acero, H.W. Hon, et al., *Spoken Language Processing*, vol. 15 (Prentice Hall, New York, 2001)

177. E. Huang, H. Zhang, D.C. Parkes, K.Z. Gajos, Y. Chen, Toward automatic task design: a progress report, in *Proceedings of the ACM SIGKDD Workshop on Human Computation* (2010), pp. 77–85

178. D. Huynh, D. Karger, Parallax and companion: set-based browsing for the data web, in *Proceedings of the 18th International Conference on World Wide Web*, WWW 2009, Madrid, Spain (ACM, New York, 2009)

179. ID3.org, ID3v2, Technical report, ID3.org. Accessed Sept 2012

180. I.F. Ilyas, W.G. Aref, A.K. Elmagarmid, Supporting top-k join queries in relational databases. VLDB J. **13**(3), 207–221 (2004)

181. I.F. Ilyas, G. Beskales, M.A. Soliman, A survey of top-k query processing techniques in relational database systems. ACM Comput. Surv. **40**(4), 11 (2008)

182. P. Ipeirotis, The new demographics of Mechanical Turk (2010), http://www.behind-the-enemy-lines.com/2010/03/new-demographics-of-mechanical-turk.html

183. P.G. Ipeirotis, P.K. Paritosh, Managing crowdsourced human computation: a tutorial, in *Proceedings of the 20th International Conference Companion on World Wide Web* (2011), pp. 287–288

184. P.G. Ipeirotis, F. Provost, J. Wang, Quality management on Amazon Mechanical Turk, in *Proceedings of the ACM SIGKDD Workshop on Human Computation (KDD-HCOMP 2010)* (2010), pp. 64–67

185. P. Jaccard, Etude comparative de la distribution florale dans une portion des alpes et des jura. Bull. Soc. Vaud. Sci. Nat. **37**(1), 547–579 (1901)

186. Japan Electronics and Information Technology Industries Association, Exchangeable image file format for digital still cameras: Exif (Version 2.2), Technical report, JEITA, 2002

187. K. Järvelin, P. Ingwersen, Information seeking research needs extension towards tasks and technology. Inf. Res. **10**(1) (2004), http://informationr.net/ir/10-1/paper212.html

188. T. Joachims, *Making Large-Scale Support Vector Machine Learning Practical* (MIT Press, Cambridge, 1999), pp. 169–184

189. A. Jøsang, R. Ismail, C. Boyd, A survey of trust and reputation systems for online service provision. Decis. Support Syst. **43**(2), 618–644 (2007)

190. A. Kalyanpur, B.K. Boguraev, S. Patwardhan, J.W. Murdock, A. Lally, C. Welty, J.M. Prager, B. Coppola, A. Fokoue-Nkoutche, L. Zhang, Y. Pan, Z.M. Qiu, Structured data and inference in DeepQA. IBM J. Res. Dev. **56**(3.4), 10 (2012). doi:10.1147/JRD.2012.2188737

191. G. Karypis, E.H. Han, V. Kumar, Chameleon: a hierarchical clustering algorithm using dynamic modeling. Computer **32**(8), 68–75 (1999)

192. G. Kasneci, F.M. Suchanek, G. Ifrim, M. Ramanath, G. Weikum, NAGA: searching and ranking knowledge, in *ICDE*, ed. by G. Alonso, J.A. Blakeley, A.L.P. Chen (IEEE Press, New York, 2008), pp. 953–962

193. D. Kempe, M. Mahdian, A cascade model for externalities in sponsored search, in *WINE* (2008), pp. 585–596

194. V. Kešelj, F. Peng, N. Cercone, C. Thomas, N-gram-based author profiles for authorship attribution, in *Proceedings of the Conference Pacific Association for Computational Linguistics*, PACLING'03 (2003)

195. R. Khare, Microformats: the next (small) thing on the semantic web? IEEE Internet Comput. **10**(1), 68–75 (2006)

196. R. Khare, T. Çelik, Microformats: a pragmatic path to the semantic web, in *WWW*, ed. by L. Carr, D.D. Roure, A. Iyengar, C.A. Goble, M. Dahlin (ACM, New York, 2006), pp. 865–866

197. F. Khatib, F. DiMaio, F.C. Group, F.V.C. Group, S. Cooper, M. Kazmierczyk, M. Gilski, S. Krzywda, H. Zabranska, I. Pichova, J. Thompson, Z. Popovic', M. Jaskolski, D. Baker, Crystal structure of a monomeric retroviral protease solved by protein folding game players. Nat. Struct. Mol. Biol. **18**(10), 1175–1177 (2011)

198. S. Kim, C. Robson, T. Zimmerman, J. Pierce, E.M. Haber, Creek watch: pairing usefulness and usability for successful citizen science, in *Proceedings of the 29th International Conference on Human Factors in Computing Systems*, New York, NY (2011), pp. 2125–2134

199. P. Kingsbury, M. Palmer, From TreeBank to PropBank, in *Proceedings of LREC* (2002)

200. A. Kiryakov, B. Popov, I. Terziev, D. Manov, D. Ognyanoff, Semantic annotation, indexing, and retrieval. J. Web Semant. **2**(1), 49–79 (2004)

201. M. Klein, The MIT deliberatorium - enabling large-scale deliberation about complex systemic problems, in *ICAART (1)*, ed. by J. Filipe, A.L.N. Fred (2011), pp. 15–24. SciTePress

202. D. Klein, C.D. Manning, Accurate unlexicalized parsing, in *Proceedings of ACL* (Association for Computational Linguistics, Stroudsburg, 2003), pp. 423–430

203. J.M. Kleinberg, Authoritative sources in a hyperlinked environment. J. ACM **46**(5), 604–632 (1999)

204. J.C. Klensin, *Role of the domain name system (DNS)*, Internet RFC 3467, Feb 2003

205. S. Koelstra, A. Yazdani, M. Soleymani, C. Mühl, J.-S. Lee, A. Nijholt, T. Pun, T. Ebrahimi, I. Patras, Single trial classification of EEG and peripheral physiological signals for recognition of emotions induced by music videos, in *Proceedings of the 2010 International Conference on Brain Informatics* (2010), pp. 89–100

206. I. Koprinska, S. Carrato, Temporal video segmentation: a survey. Signal Process. Image Commun. **16**(5), 477–500 (2001)

207. M. Koster, *A method for web robots control*, Internet Draft draft-koster-robots-00, Dec (1996)

208. G. Koutrika, Y. Ioannidis, Personalized queries under a generalized preference model, in *Proceedings of the 21st International Conference on Data Engineering, 2005*, ICDE 2005 (2005), pp. 841–852. doi:10.1109/ICDE.2005.106

209. C.C. Kuhlthau, Inside the search process: information seeking from the user's perspective. J. Am. Soc. Inf. Sci. **42**(5), 361–371 (1991)

210. C.C. Kuhlthau, *Seeking Meaning: a Process Approach to Library and Information Services/Carol Collier Kuhlthau* (Ablex, Norwood, 1993), p. 199
211. L.I. Kuncheva, C.J. Whitaker, C.A. Shipp, Limits on the majority vote accuracy in classifier fusion. Pattern Anal. Appl. **6**(1), 22–31 (2003)
212. J.D. Lafferty, A. McCallum, F.C.N. Pereira, Conditional random fields: probabilistic models for segmenting and labeling sequence data, in *ICML*, ed. by C.E. Brodley, A.P. Danyluk (Morgan Kaufmann, San Mateo, 2001), pp. 282–289
213. F.W. Lancaster, E.G. Fayen, *Information Retrieval On-Line* (Melville, New York, 1973)
214. K. Lang, J. Delgado, D. Jiang, B. Ghosh, S. Das, A. Gajewar, S. Jagadish, A. Seshan, C. Botev, M. Bindeberger-Ortega, et al., Efficient online ad serving in a display advertising exchange, in *ACM WSDM* (ACM, New York, 2011), pp. 307–316
215. A. Langville, C. Meyer, *Google's Page Rank and Beyond: the Science of Search Engine Rankings* (Princeton University Press, Princeton, 2008)
216. O. Lassila, R.R. Swick, Resource description framework (RDF) model and syntax specification, Technical report, W3C Recommendation, Feb 1999, http://www.w3.org/TR/1999/REC-rdf-syntax-19990222/
217. N. Lathia, S. Hailes, L. Capra, X. Amatriain, Temporal diversity in recommender systems, in *Proceedings of the 33rd International ACM SIGIR Conference on Research and Development in Information Retrieval*. SIGIR'10 (ACM, New York, 2010), pp. 210–217
218. M. Lebowitz, Experiments with incremental concept formation. Mach. Learn. **2**, 103–138 (1987)
219. K.-F. Lee, *Automatic Speech Recognition: the Development of the Sphinx Recognition System*, vol. 62 (Kluwer Academic, Norwell, 1989)
220. C. Lee, Y.-G. Hwang, M.-G. Jang, Fine-grained named entity recognition and relation extraction for question answering, in *Proceedings of the 30th Annual International ACM SIGIR Conference on Research and Development in Information Retrieval*. SIGIR'07 (ACM, New York, 2007), pp. 799–800
221. R. Lempel, S. Moran, Predictive caching and prefetching of query results in search engines, in *Proceedings of the 12th International Conference on World Wide Web*. WWW'03 (ACM, New York, 2003), pp. 19–28
222. R. Lempel, S. Moran, Rank-stability and rank-similarity of link-based web ranking algorithms in authority-connected graphs. Inf. Retr. **8**(2), 245–264 (2005)
223. N. Lester, A. Moffat, J. Zobel, Fast on-line index construction by geometric partitioning, in *Proceedings of the 14th ACM International Conference on Information and Knowledge Management*. CIKM'05 (ACM, New York, 2005), pp. 776–783
224. M.S. Lew, N. Sebe, C. Djeraba, R. Jain, Content-based multimedia information retrieval: state of the art and challenges. ACM Trans. Multimed. Comput. Commun. Appl. **2**(1), 1–19 (2006)
225. X. Li, D. Roth, Learning question classifiers, in *Proceedings of the 19th International Conference on Computational Linguistics—Volume 1*. COLING'02 (Association for Computational Linguistics, Stroudsburg, 2002), pp. 1–7
226. Limelight Networks, Limelight video platform, http://www.limelightvideoplatform.com/. Accessed Sept 2012
227. X. Ling, D.S. Weld, Temporal information extraction, in *Proceedings of the Twenty Fifth National Conference on Artificial Intelligence* (2010)
228. Y. Liu, D. Zhang, G. Lu, W.-Y. Ma, A survey of content-based image retrieval with high-level semantics. Pattern Recognit. **40**(1), 262–282 (2007)
229. Z. Liu, P. Sun, Y. Chen, Structured search result differentiation. Proc. VLDB Endow. **2**(1), 313–324 (2009)
230. Y. Liu, P. Piyawongwisal, S. Handa, L. Yu, Y. Xu, A. Samuel, Going beyond citizen data collection with Mapster: a Mobile+Cloud citizen science experiment, in *Proceedings of the IEEE Workshop on Computing for Citizen Science (eScience 2001)* (2011)
231. D.G. Lowe, Distinctive image features from scale-invariant keypoints. Int. J. Comput. Vis. **60**(2), 91–110 (2004)

232. H.P. Luhn, A statistical approach to mechanized encoding and searching of literary information. IBM J. Res. Dev. **1**(4), 309–317 (1957)
233. H.P. Luhn, The automatic creation of literature abstracts. IBM J. Res. Dev. **2**(2), 159–165 (1958)
234. J. Macqueen, Some methods for classification and analysis of multivariate observations, in *Proceedings of the 5th Berkeley Symposium on Mathematical Statistics and Probability*, vol. 1 (University of California Press, Berkeley, 1967), pp. 281–297
235. A. Madan, M. Cebrián, D. Lazer, A. Pentland, Social sensing for epidemiological behavior change, in *Proceedings of the 12th ACM International Conference on Ubiquitous Computing* (2010), pp. 291–300
236. J. Madhavan, D. Ko, L. Kot, V. Ganapathy, A. Rasmussen, A.Y. Halevy, Google's deep web crawl. Proc. VLDB Endow. **1**(2), 1241–1252 (2008)
237. N. Maisonneuve, M. Stevens, M.E. Niessen, P. Hanappe, L. Steels, Citizen noise pollution monitoring, in *Proceedings of the 10th Annual International Conference on Digital Government Research: Social Networks: Making Connections Between Citizens, Data and Government* (2009), pp. 96–103
238. C. Manasseh, K. Ahern, R. Sengupta, The connected traveler: using location and personalization on mobile devices to improve transportation, in *Proceedings of the 2nd International Workshop on Location and the Web (LOCWEB09)* (2009), pp. 1–4
239. I. Mani, G. Wilson, Robust temporal processing of news, in *Proceedings of the 38th Annual Meeting on Association for Computational Linguistics* (Association for Computational Linguistics, Stroudsburg, 2000), pp. 69–76
240. B.S. Manjunath, P. Salembier, T. Sikora, *Introduction to MPEG-7: Multimedia Content Description Interface* (Wiley, New York, 2002)
241. C.D. Manning, P. Raghavan, H. Schütze, Introduction to information retrieval. 2008. Online edition (2007)
242. P. Maragos, A. Potamianos, P. Gros, *Multimodal Processing and Interaction, Audio, Video, Text Series: Multimedia Systems and Applications*, vol. 33, 1st edn. (Springer, Berlin, 2008)
243. G. Marchionini, *Information Seeking in Electronic Environments* (Cambridge University Press, New York, 1995)
244. G. Marchionini, Exploratory search: from finding to understanding. Commun. ACM **49**(4), 41–46 (2006)
245. E.P. Markatos, On caching search engine query results, in *Computer Communications* (2000)
246. A. Marsden, A. Mackenzie, A. Lindsay, H. Nock, J. Coleman, G. Kochanski, Tools for searching, annotation and analysis of speech, music, film and video—a survey. Lit. Linguist. Comput. **22**(4), 469–488 (2007)
247. D. Martinenghi, M. Tagliasacchi, Top-*k* pipe join, in *ICDE Workshops* (IEEE Press, San Diego, 2010), pp. 16–19
248. J.M. Martínez, *MPEG-7 overview (version 10)*, ISO/IEC JTC1/SC29/WG11N6828, 2004
249. A. McCallum, MALLET: a machine learning for language toolkit, http://mallet.cs.umass.edu/. Accessed Sept 2012
250. S. McCarron, XHTML+RDFa 1.1—support for RDFa via XHTML modularization, Technical report, W3C Recommendatio, June 2012, http://www.w3.org/TR/xhtml-rdfa/
251. D.L. McGuinness, F. Van Harmelen, OWL web ontology language overview, Technical report, W3C Recommendation, Nov 2004, http://www.w3.org/TR/owl-features/
252. E. Meij, M. Bron, L. Hollink, B. Huurnink, M. de Rijke, Mapping queries to the linking open data cloud: a case study using DBpedia. J. Web Semant. **9**(4), 418–433 (2011)
253. S. Melink, S. Raghavan, B. Yang, H. Garcia-Molina, Building a distributed full-text index for the web. ACM Trans. Inf. Syst. **19**(3), 217–241 (2001)
254. G.A. Miller, WordNet: a lexical database for English. Commun. ACM **38**(11), 39–41 (1995)
255. A. Moffat, W. Webber, J. Zobel, R. Baeza-Yates, A pipelined architecture for distributed text query evaluation. Inf. Retr. **10**(3), 205–231 (2007)
256. A. Moschitti, S. Quarteroni, Linguistic kernels for answer re-ranking in question answering systems. Inf. Process. Manag. **47**(6), 825–842 (2011)

257. A. Moschitti, S. Quarteroni, R. Basili, S. Manandhar, Exploiting syntactic and shallow semantic kernels for question answer classification, in *ACL* (2007)
258. M. Mozer, H. Pashler, M.H. Wilder, R.A. Lindsey, M. Jones, M. Jones, Improving human judgments by decontaminating sequential dependencies, in *Advances in Neural Information Processing Systems* (2010), pp. 1705–1713
259. H. Mühleisen, C. Bizer, Web data commons—extracting structured data from two large web corpora, in *LDOW*, ed. by C. Bizer, T. Heath, T. Berners-Lee, M. Hausenblas. CEUR Workshop Proceedings, vol. 937 (2012). CEUR-WS.org
260. S. Muthukrishnan, Ad exchanges: research issues, in *WINE* (2009), pp. 1–12
261. M. Najork, Web crawler architecture, in *Encyclopedia of Database Systems*, ed. by L. Liu, M.T. Öñzsu (Springer, Berlin, 2009), pp. 3462–3465
262. M. Najork, J.L. Wiener, Breadth-first crawling yields high-quality pages, in *Proceedings of the 10th International Conference on World Wide Web*. WWW'01 (ACM, New York, 2001), pp. 114–118
263. M.A. Najork, H. Zaragoza, M.J. Taylor, Hits on the web: how does it compare? in *Proceedings of the 30th Annual International ACM SIGIR Conference on Research and Development in Information Retrieval*. SIGIR'07 (ACM, New York, 2007), pp. 471–478
264. Y. Narahari, D. Garg, R. Narayanam, H. Prakash, *Game Theoretic Problems in Network Economics and Mechanism Design Solutions* (Springer, Berlin, 2009)
265. F. Naumann, M. Herschel, *An Introduction to Duplicate Detection*. Synthesis Lectures on Data Management (Morgan & Claypool, San Rafael, 2010)
266. A. Nenkova, K. McKeown, *Automatic Summarization* (Now Publishers, Hanover, 2011)
267. C.-W. Ngo, C.-K. Chan, Video text detection and segmentation for optical character recognition. Multimed. Syst. **10**, 261–272 (2005)
268. Nielsen, What Americans do online: social media and games dominate activity, Technical report, Nielsen, Aug 2011
269. H.N. Njuguna, Are smart phones better than paper-based questionnaires for surveillance data collection? A comparative evaluation using influenza sentinel surveillance sites in Kenya, 2011, in *Proc. International Conference on Emerging Infectious Deseases*, ed. by A.S. for Microbiology. ICEID 2012, Mar 2012, (2012)
270. D.W. Oard, D. He, J. Wang, User-assisted query translation for interactive cross-language information retrieval. Inf. Process. Manag. **44**(1), 181–211 (2008)
271. L. Page, S. Brin, R. Motwani, T. Winograd, The PageRank citation ranking: bringing order to the web, Technical report, Stanford InfoLab, 1999
272. G. Pandurangan, P. Raghavan, E. Upfal, Using PageRank to characterize web structure, in *Proceedings of the 8th Annual International Conference on Computing and Combinatorics*. COCOON'02 (Springer, London, 2002), pp. 330–339
273. A. Parameswaran, A.D. Sarma, H. Garcia-Molina, N. Polyzotis, J. Widom, Human-assisted graph search: it's okay to ask questions, Technical report, Stanford University, Palo Alto, CA, 2010
274. D.M.R. Park, Concurrency and automata on infinite sequences, in *Theoretical Computer Science*, ed. by P. Deussen. Lecture Notes in Computer Science, vol. 104 (Springer, Berlin, 1981), pp. 167–183
275. D. Petrovska-Delacrétaz, A. El Hannani, G. Chollet, Text-independent speaker verification: state of the art and challenges, in *Progress in Nonlinear Speech Processing*, ed. by Y. Stylianou, M. Faundez-Zanuy, A. Esposito. Lecture Notes in Computer Science, vol. 4391 (Springer, Berlin, 2007), pp. 135–169
276. G. Pickard, I. Rahwan, W. Pan, M. Cebrián, R. Crane, A. Madan, A. Pentland, Time critical social mobilization: the DARPA network challenge winning strategy, http://www.sciencemag.org/content/334/6055/509 (2010)
277. P.L.T. Pirolli, *Information Foraging Theory: Adaptive Interaction with Information*, 1st edn. (Oxford University Press, New York, 2007)
278. J. Plisson, N. Lavrac, D. Mladenić, A rule based approach to word lemmatization. Knowledge, 83–86 (2004), http://eprints.pascal-network.org/archive/00000715/

279. A. Poggi, D. Lembo, D. Calvanese, G.D. Giacomo, M. Lenzerini, R. Rosati, Linking data to ontologies. J. Data Semant. **10**, 133–173 (2008)

280. M.F. Porter, An algorithm for suffix stripping. Program, Electron. Libr. Inf. Syst. **40**(3), 211–218 (1980)

281. I. Potamitis, T. Ganchev, Generalized recognition of sound events: approaches and applications, in *Multimedia Services in Intelligent Environments*, ed. by G. Tsihrintzis, L. Jain. Studies in Computational Intelligence, vol. 120 (Springer, Berlin, 2008), pp. 41–79

282. S. Quarteroni, Question answering, semantic search and data service querying, in *Proceedings of the KRAQ11 Workshop* (Asian Federation of Natural Language Processing, Chiang Mai, 2011), pp. 10–17

283. S. Quarteroni, S. Manandhar, Designing an interactive open-domain question answering system. Nat. Lang. Eng. **15**(1), 73–95 (2009)

284. S. Quarteroni, A.V. Ivanov, G. Riccardi, Simultaneous dialog act segmentation and classification from human–human spoken conversations, in *2011 IEEE International Conference on Acoustics, Speech and Signal Processing (ICASSP)*, (IEEE Press, New York, 2011), pp. 5596–5599

285. S. Quarteroni, M. Brambilla, S. Ceri, A bottom-up, knowledge-aware approach to the integration of web data services, in *ACM-TWEB*. To appear

286. J.R. Quinlan, *C4.5: Programs for Machine Learning* (Morgan Kaufmann, San Mateo, 1993)

287. A.J. Quinn, B.B. Bederson, Human computation: a survey and taxonomy of a growing field, in *Proceedings of the 29th International Conference on Human Factors in Computing Systems* (2011), pp. 1403–1412

288. L.R. Rabiner, A tutorial on hidden Markov models and selected applications in speech recognition. Proc. IEEE **77**(2), 257–286 (1989)

289. R.J. Radke, S. Andra, O. Al-Kofahi, B. Roysam, Image change detection algorithms: a systematic survey. IEEE Trans. Image Process. **14**(3), 294–307 (2005)

290. F. Radlinski, S. Dumais, Improving personalized web search using result diversification, in *Proceedings of the 29th Annual International ACM SIGIR Conference on Research and Development in Information Retrieval*. SIGIR'06 (ACM, New York, 2006), pp. 691–692

291. D. Rafiei, K. Bharat, A. Shukla, Diversifying web search results, in *WWW'10: Proceedings of the 19th International Conference on World Wide Web* (ACM, New York, 2010), pp. 781–790

292. A. Rajaraman, Kosmix: high-performance topic exploration using the deep web. Proc. VLDB Endow. **2**(2), 1524–1529 (2009)

293. Z. Rasheed, M. Shah, Scene detection in Hollywood movies and TV shows, in *Proceedings of the 2003 IEEE Computer Society Conference on Computer Vision and Pattern Recognition*, vol. 2 (2003), pp. 343–348

294. B. Ribeiro-Neto, E.S. Moura, M.S. Neubert, N. Ziviani, Efficient distributed algorithms to build inverted files, in *Proceedings of the 22nd Annual International ACM SIGIR Conference on Research and Development in Information Retrieval*. SIGIR'99 (ACM, New York, 1999), pp. 105–112

295. M. Richardson, A. Prakash, E. Brill, Beyond PageRank: machine learning for static ranking, in *Proceedings of the 15th International Conference on World Wide Web*. WWW'06 (ACM, New York, 2006), pp. 707–715

296. S.E. Robertson, S. Walker, Okapi/keenbow at trec-8, in *Proc. of TREC*, vol. 8 (1999)

297. P.M. Roget, J.L. Roget, S.R. Roget, *Thesaurus of English Words and Phrases: Classified and Arranged so as to Facilitate the Expression of Ideas and Assist in Literary Composition* (Longmans, Green, New York, 1960)

298. D.E. Rose, The information-seeking funnel, in *National Science Foundation Workshop on Information-Seeking Support Systems (ISSS)*, ed. by G. Marchionini, R. White, Chapel Hill, NC, June (2008)

299. P.J. Rousseeuw, Silhouettes: a graphical aid to the interpretation and validation of cluster analysis. J. Comput. Appl. Math. **20**, 53–65 (1987)

300. T. Sakaki, M. Okazaki, Y. Matsuo, Earthquake shakes Twitter users: real-time event detection by social sensors, in *Proceedings of the 19th International Conference on World Wide Web (WWW10)* (2010), pp. 851–860
301. G. Salton, *Automatic Information Organization and Retrieval* (McGraw-Hill, New York, 1968)
302. G. Salton, C. Buckley, Term-weighting approaches in automatic text retrieval. Inf. Process. Manag. **24**(5), 513–523 (1988)
303. G. Salton, C. Buckley, Improving retrieval performance by relevance feedback, in *Readings in Information Retrieval* (1997), pp. 355–364
304. G. Salton, A. Wong, C.S. Yang, A vector space model for automatic indexing. Commun. ACM **18**, 613–620 (1975)
305. A.G. Sanfey, Social decision-making: insights from game theory and neuroscience. Science **318**(5850), 598–602 (2007)
306. E. Saquete, P. Martinez-Barco, R. Munoz, J. Vicedo, Splitting complex temporal questions for question answering systems, in *Proceedings of the 42nd Annual Meeting on Association for Computational Linguistics* (Association for Computational Linguistics, Stroudsburg, 2004), p. 566
307. P.C. Saraiva, E. Silva de Moura, N. Ziviani, W. Meira, R. Fonseca, B. Riberio-Neto, Rank-preserving two-level caching for scalable search engines, in *Proceedings of the 24th Annual International ACM SIGIR Conference on Research and Development in Information Retrieval*. SIGIR'01 (ACM, New York, 2001), pp. 51–58
308. N. Scaringella, G. Zoia, D. Mlynek, Automatic genre classification of music content: a survey. IEEE Signal Process. Mag. **23**(2), 133–141 (2006)
309. J.B. Schafer, J.A. Konstan, J. Riedl, E-commerce recommendation applications. Data Min. Knowl. Discov. **5**(1-2), 115–153 (2001)
310. M.M.C. Schraefel, M. Wilson, A. Russell, D.A. Smith, mSpace: improving information access to multimedia domains with multimodal exploratory search. Commun. ACM **49**(4), 47–49 (2006)
311. F. Sebastiani, Machine learning in automated text categorization. ACM Comput. Surv. **34**(1), 1–47 (2002)
312. S. Shah, F. Bao, C.T. Lu, I.R. Chen, CROWDSAFE: crowd sourcing of crime incidents and safe routing on mobile devices, in *Proceedings of the 19th ACM SIGSPATIAL International Conference on Advances in Geographic Information Systems* (ACM, New York, 2011), pp. 521–524
313. P. Sheridan, M. Braschlert, P. Schäuble, Cross-language information retrieval in a multilingual legal domain, in *Research and Advanced Technology for Digital Libraries* (1997), pp. 253–268
314. C.C. Shilakes, J. Tylman, *Enterprise Information Portals* (Merrill, Columbus, 1998), p. 16
315. B. Shneiderman, D. Byrd, W.B. Croft, Clarifying search: a user-interface framework for text searches. D-Lib Mag. **3**(1) (1997)
316. R.F. Simmons, Answering English questions by computer: a survey. Commun. ACM **8**(1), 53–70 (1965)
317. D. Skoutas, M. Alrifai, W. Nejdl, Re-ranking web service search results under diverse user preferences, in *PersDB 2010*, September (2010)
318. C.G.M. Snoek, M. Worring, Concept-based video retrieval. Found. Trends Inf. Retr. **2**(4), 215–322 (2009)
319. T. Steiner, L. Sutton, S. Spiller, M. Lazzaro, F.S. Nucci, V. Croce, A. Massari, A. Camurri, A. Verroust-Blondet, L. Joyeux, J. Etzold, P. Grimm, A. Mademlis, S. Malassiotis, P. Daras, A. Axenopoulos, D. Tzovaras, I-SEARCH: a multimodal search engine based on rich unified content description (RUCoD), in *WWW (Companion Volume)*, ed. by A. Mille, F.L. Gandon, J. Misselis, M. Rabinovich, S. Staab (ACM, New York, 2012), pp. 291–294
320. W. Stewart, *Introduction to the Numerical Solution of Markov Chains* (Princeton University Press, Princeton, 1994)

321. A. Stolcke, SRILM-an extensible language modeling toolkit, in *Seventh International Conference on Spoken Language Processing* (2002)
322. J.R. Stothard, J.C. Sousa-Figueiredo, M. Betson, E.Y.W. Seto, N.B. Kabatereine, Investigating the spatial micro-epidemiology of diseases within a point-prevalence sample: a field applicable method for rapid mapping of households using low-cost gps-dataloggers. Trans. R. Soc. Trop. Med. Hyg. (2011), http://trstmh.oxfordjournals.org/content/105/9/500.short
323. F.M. Suchanek, G. Kasneci, G. Weikum, Yago: a large ontology from Wikipedia and WordNet. J. Web Semant. **6**(3), 203–217 (2008)
324. Y. Sun, Z. Zhuang, C.L. Giles, A large-scale study of robots.txt, in *WWW*, ed. by C.L. Williamson, M.E. Zurko, P.F. Patel-Schneider, P.J. Shenoy (ACM, New York, 2007), pp. 1123–1124
325. M. Surdeanu, M. Ciaramita, H. Zaragoza, Learning to rank answers to non-factoid questions from web collections. Comput. Linguist. **37**(2), 351–383 (2011)
326. The Quaero Consortium, The Quaero program, http://www.quaero.org. Accessed Sept 2012
327. Theseus, The Theseus programme, http://theseus-programm.de. Accessed Sept 2012
328. R. Tobin, Barriers on a search results page, Technical report, Enquiro Research, 2008
329. A. Tomasic, H. García-Molina, K. Shoens, Incremental updates of inverted lists for text document retrieval. SIGMOD Rec. **23**(2), 289–300 (1994)
330. T. Tran, D.M. Herzig, G. Ladwig, SemSearchPro—using semantics throughout the search process. J. Web Semant. **9**(4), 349–364 (2011)
331. D. Tunkelang, *Faceted Search*. Synthesis Lectures on Information Concepts, Retrieval, and Services (Morgan & Claypool Publishers, San Rafael, 2009)
332. P.K. Turaga, R. Chellappa, V.S. Subrahmanian, O. Udrea, Machine recognition of human activities: a survey. IEEE Trans. Circuits Syst. Video Technol. **18**(11), 1473–1488 (2008)
333. A. Tversky, D. Kahneman, Judgment under uncertainty: heuristics and biases. Science **185**, 1124–1131 (1974)
334. R. Typke, F. Wiering, R.C. Veltkamp, A survey of music information retrieval systems, in *Proceedings ISMIR 2005, 6th International Conference on Music Information Retrieval*, London, UK, 11–15 September (2005), pp. 153–160
335. C. Tzviskou, M. Brambilla, Semantic personalization of web portal contents, in *Proceedings of the 16th International Conference on World Wide Web*. WWW'07 (ACM, New York, 2007), pp. 1245–1246
336. University of Twente, Buchenwald demonstrator, http://vuurvink.ewi.utwente.nl. Accessed Sept 2012
337. T. Upstill, et al., Predicting fame and fortune: PageRank or Indegree? in *In Proceedings of the Australasian Document Computing Symposium*, ADCS 2003 (2003), pp. 31–40
338. V.S. Uren, Y. Lei, V. Lopez, H. Liu, E. Motta, M. Giordanino, The usability of semantic search tools: a review. Knowl. Eng. Rev. **22**(4), 361–377 (2007)
339. B. Uzzi, The sources and consequences of embeddedness for the economic performance of organizations: the network effect. Am. Sociol. Rev. **61**(4), 674–698 (1996)
340. V.N. Vapnik, *The Nature of Statistical Learning Theory* (Springer, New York, 2000)
341. H.R. Varian, Position auctions. Int. J. Ind. Organ. **25**(6), 1163–1178 (2007)
342. E. Vee, U. Srivastava, J. Shanmugasundaram, P. Bhat, S.A. Yahia, Efficient computation of diverse query results, in *ICDE'08: Proceedings of the 2008 IEEE 24th International Conference on Data Engineering* (IEEE Comput. Soc., Washington, 2008), pp. 228–236
343. L. von Ahn, Human computation, in *CIVR* (2009)
344. L. von Ahn, Human computation, Ph.D. thesis, CMU, CMU-CS-05-193, Dec 2005
345. L. von Ahn, L. Dabbish, Labeling images with a computer game, in *CHI*, ed. by E. Dykstra-Erickson, M. Tscheligi (ACM, New York, 2004), pp. 319–326
346. L. von Ahn, L. Dabbish, Designing games with a purpose. Commun. ACM **51**(8), 58–67 (2008)
347. L. von Ahn, R. Liu, M. Blum, Peekaboom: a game for locating objects in images, in *Proceedings of the SIGCHI Conference on Human Factors in Computing Systems*. CHI'06 (ACM, New York, 2006), pp. 55–64

348. L. von Ahn, B. Maurer, C. McMillen, D. Abraham, M. Blum, reCAPTCHA: human-based character recognition via web security measures. Science **321**(5895), 1465–1468 (2008)

349. R.W. White, R.A. Roth, *Exploratory Search: Beyond the Query-Response Paradigm*. Synthesis Lectures on Information Concepts, Retrieval, and Services (Morgan & Claypool Publishers, San Rafael, 2009)

350. R.W. White, J.M. Jose, I. Ruthven, A task-oriented study on the influencing effects of query-biased summarisation in web searching. Inf. Process. Manag. **39**(5), 707–733 (2003)

351. M.L. Wilson, *Search User Interface Design*. Synthesis Lectures on Information Concepts, Retrieval, and Services (Morgan & Claypool Publishers, San Rafael, 2011)

352. D.R. Wilson, T.R. Martinez, Improved heterogeneous distance functions. J. Artif. Intell. Res. **6**, 1–34 (1997)

353. I.H. Witten, G.W. Paynter, E. Frank, C. Gutwin, C.G. Nevill-Manning, KEA: practical automatic keyphrase extraction, in *Proceedings of the Fourth ACM Conference on Digital Libraries*, (ACM, New York, 1999), pp. 254–255

354. I.H. Witten, E. Frank, M.A. Hall, *Data Mining: Practical Machine Learning Tools and Techniques*, 3rd edn. (Morgan Kaufmann, San Mateo, 2011)

355. Q. Wu, D.X. Zhou, SVM soft margin classifiers: linear programming versus quadratic programming. Neural Comput. **17**(5), 1160–1187 (2005)

356. Y. Xie, D. O'Hallaron, Locality in search engine queries and its implications for caching, in *IEEE Infocom 2002* (2002), pp. 1238–1247

357. T. Yan, V. Kumar, D. Ganesan, CrowdSearch: exploiting crowds for accurate real-time image search on mobile phones, in *MobiSys*, ed. by S. Banerjee, S. Keshav, A. Wolman (ACM, New York, 2010), pp. 77–90

358. Y. Yang, An evaluation of statistical approaches to text categorization. Inf. Retr. **1**(1), 69–90 (1999)

359. Z. Yang, B. Li, Y. Zhu, I. King, G.-A. Levow, H.M. Meng, Collection of user judgments on spoken dialog system with crowdsourcing, in *SLT*, ed. by D. Hakkani-Tür, M. Ostendorf (IEEE Press, New York, 2010), pp. 277–282

360. K.-P. Yee, K. Swearingen, K. Li, M. Hearst, Faceted metadata for image search and browsing, in *Proceedings of the SIGCHI Conference on Human Factors in Computing Systems*. CHI'03 (ACM, New York, 2003), pp. 401–408

361. T. Yeh, B. White, J. San Pedro, B. Katz, L.S. Davis, A case for query by image and text content: searching computer help using screenshots and keywords, in *Proceedings of the 20th International Conference on World Wide Web*. WWW'11 (ACM, New York, 2011), pp. 775–784

362. A. Yilmaz, O. Javed, M. Shah, Object tracking: a survey. ACM Comput. Surv. **38**(4), 13 (2006)

363. M.A. Yosef, J. Hoffart, I. Bordino, M. Spaniol, G. Weikum, Aida: an online tool for accurate disambiguation of named entities in text and tables. Proc. VLDB Endow. **4**(12), 1450–1453 (2011)

364. G.-J. Yu, Y.-S. Chen, K.-P. Shih, A content-based image retrieval system for outdoor ecology learning: a firefly watching system, in *AINA 2* (IEEE Comput. Soc., Los Alamitos, 2004), pp. 112–115

365. J. Yu, B. Benatallah, F. Casati, F. Daniel, Understanding mashup development. IEEE Internet Comput. **12**(5), 44–52 (2008)

366. J.X. Yu, L. Qin, L. Chang, Keyword search in relational databases: a survey. IEEE Data Eng. Bull. **33**(1), 67–78 (2010)

367. D. Zelenko, C. Aone, A. Richardella, Kernel methods for relation extraction, in *JMLR* (2003)

368. C. Zhai, J. Lafferty, A study of smoothing methods for language models applied to ad hoc information retrieval, in *Proceedings of the 24th Annual International ACM SIGIR Conference on Research and Development in Information Retrieval*. SIGIR'01 (ACM, New York, 2001), pp. 334–342

369. D. Zhang, W.S. Lee, Question classification using support vector machines, in *Proceedings of SIGIR* (ACM, New York, 2003)

370. T. Zhang, R. Ramakrishnan, M. Livny, BIRCH: an efficient data clustering method for very large databases, in *SIGMOD Conference* (1996)

371. W.-Y. Zhao, R. Chellappa, P.J. Phillips, A. Rosenfeld, Face recognition: a literature survey. ACM Comput. Surv. **35**(4), 399–458 (2003)

372. G.D. Zhou, J. Su, Named entity recognition using an HMM-based chunk tagger, in *Proceedings of the 40th Annual Meeting on Association for Computational Linguistics* (Association for Computational Linguistics, Stroudsburg, 2002), pp. 473–480

373. K.G. Zipf, *Human Behavior and the Principle of Least Effort* (Addison-Wesley, Reading, 1949)

374. J. Zittrain, Ubiquitous human computing. Philos. Trans. R. Soc., Math. Phys. Eng. Sci. **366**(1881), 3813–3821 (2008)

Index

S. Ceri et al., *Web Information Retrieval*, Data-Centric Systems and Applications,
DOI 10.1007/978-3-642-39314-3, © Springer-Verlag Berlin Heidelberg 2013